AN INTRODUCTION TO FUNCTIONS THROUGH APPLICATIONS

INTERMEDIATE ALGEBRA

NKATE CURRICULUM PROJECT
CLASS TEST EDITION

NATIONAL · SCIENCE · FOUNDATION

▲▲ ADDISON-WESLEY

An imprint of Addison Wesley Longman, Inc.

Reading, Massachusetts • Menlo Park, California • New York • Harlow, England
Don Mills, Ontario • Sydney • Mexico City • Madrid • Amsterdam

The chameleon graphic associated with the NKATE project was derived from the Papago symbol for "change and the courage it takes." The cover concept for this class test edition is by Kathy Mowers and Steven Paulo Davis, the electronic design and execution by Holly Bowyer.

The following material from *Direct and Inverse Variation* (© July 1993) is included in this class test edition courtesy of author Kathy Mowers. Copyright 1993, Kathy Mowers. Reprinted with permission.

p. 48	Electricity Usage
p. 49	*You Try It*
p. 122	Inverse Variation
p. 124	Oxygen Tank
p. 125	*You Try It* (Peach, Inc.)
p. 138	Problems 41 and 42

This book was produced using Ami Pro 3.1™ by Lotus Development Corp.

m&m's® are a registered trademark of m&m/Mars a division of Mars, Inc., Hackettstown, NJ 07840-9150

Reproduced by Addison Wesley Longman from camera-ready copy supplied by the author.

ISBN 0-201-85361-2
1 2 3 4 5 6 7 8 9 10 VG 99989796

University of Kentucky NKATE Curriculum Project Faculty

Darrell H. Abney
Maysville Community College

Eric Allison
Elizabethtown Community College

Roger Angevine
Somerset Community College

Dana T. Calland
Maysville Community College

Karin Chess
Owensboro Community College

Maura Corley
Henderson Community College

Lillie R. F. Crowley
Lexington Community College

Vince DiNoto
Jefferson Community College

Billy Dobbs
Somerset Community College

Brita Dockstader
Jefferson Community College

Sharon Griggs
St. Petersburg Junior College

Rebecca Isaac-Fahey
Lexington Community College

Lynn Molloy
Lexington Community College

Kathy R. Mowers
Owensboro Community College

Terry Pasley
Maysville Community College

Jim Stewart
Jefferson Community College

Mike Stewart
Paducah Community College

Roger Warren
Madisonville Community College

Partial support for this work was provided by the National Science Foundation Advanced Technological Education Program through grant DUE-9454585. Any opinions, findings, conclusions, or recommendations expressed herein are those of the authors and do not necessarily reflect the views of NSF.

Table of Contents

Preface to the Instructor

Perhaps the biggest problem facing college mathematics programs across the nation is the number of entering students who must repeat high school level mathematics courses before attempting college-level mathematics. This is particularly serious in two-year colleges where more than half of entering students begin with a high school level algebra or prealgebra course. High withdrawal rates in these courses compound the problem. Success rates of 50% or less mean few of our students survive to study calculus, science, or technology courses. Additionally, the courses we offer are generally very traditional drills-and-skills courses taught in a lecture format. Uri Treisman has said that in our algebra courses we are saying the same things students have heard (and not understood) for years; we're just saying them faster and louder, and hoping they will "catch it" this time.

Our desire to change this situation led to the development of this text. The chameleon on the cover symbolizes our desire to change the way we teach. Although written for intermediate algebra, it is not a typical algebra book. We have developed a text with essentially the same goals as the traditional intermediate algebra book using a different approach and delivery system. Application problems, presented using technology and collaborative learning, motivate topics. We approach each topic graphically, numerically, and algebraically, and we have written the text mindful of the AMATYC *Standards* and the calculus reform movement. We want students to be able to communicate the mathematics they learn, and we expect this approach to better prepare them for further study in mathematics, science, and technology.

You will not find all the topics included in a typical intermediate algebra text, but how many of us really had time to cover them anyway? One of the changes adopted is the philosophy that "less is best," and many mathematical dinosaurs are omitted to allow time to cover topics in more depth. Our two major objectives are to reduce significantly the number of students who traditionally are not successful at this level and to increase student success at the next level.

Whenever possible, we have experiments or "hands-on" projects to introduce or reinforce topics. Application problems are included to answer the question, "Why do I have to study this?" The instructor becomes a facilitator who will coordinate classroom activities, not merely disperse information by lecture.

A list of elementary algebra skills that we feel are prerequisite to the study of *An Introduction to Functions Through Applications* follows the student preface. Be prepared to give references to students who need to brush up on some of these topics.

Symbols are used throughout the book to assist you in your assignments. The ✎ marks areas for student interaction. The ⌘ marks a problem designed to be worked in groups of 3 or 4 students in a collaborative learning environment. The ✔ indicates review exercises.

As William Butler Yates said, "Education is not the filling of the pail; it is the lighting of the fire." We must change our methods to prepare our students mathematically for the technological world. Remember, we cannot just talk louder and faster, continuing to teach as before.

Preface to the Student

As you read the Preface to the Instructor, you will see that this text proposes to be different from the typical intermediate algebra book. The chameleon symbolizes our commitment to change the way you learn mathematics. Just what changes do we propose?

This text uses experiments and "hands-on" projects to introduce and reinforce topics. Application problems are included to answer the age-old question, "Why do I have to study this?" Calculator technology is used wherever appropriate to reinforce algebraic topics or to replace tedious manipulations.

A list of essential skills needed to study *An Introduction to Functions Through Applications* is included on the following page. If you encounter difficulty with any of these topics, ask your instructor for a reference that covers the topic in detail.

We hope you enjoy the course as much as the students who were in previous pilot sessions. The course will be demanding but far more rewarding than a traditional course. Good luck!

Prerequisite Skills

In order to study *An Introduction to Functions Through Applications* you should be able to:

1. Simplify algebraic expressions by removing grouping symbols and combining like terms;

2. Solve linear equations in one variable;

3. Solve literal equations for the indicated variable;

4. Simplify expressions using rules of exponents;

5. Express numbers in scientific notation;

6. Factor expressions involving the greatest common factor, binomials, and quadratic trinomials;

7. Graph linear equations using a table of values;

8. Solve linear inequalities in one variable; and

9. Find the area and perimeter or circumference of triangles, rectangles, squares, and circles.

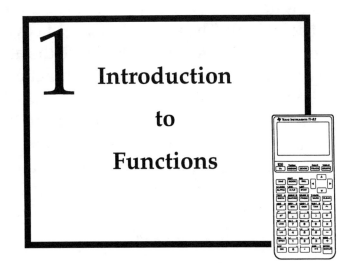

1 Introduction to Functions

Upon successful completion of this unit you should be able to:

1. Supply at least three reasons for studying functions;

2. Describe at least three applications of functions from everyday life;

3. Relate applications of functions in your major area;

4. State four methods of representing functions; and

5. Use your calculator to evaluate functions numerically.

Functions provide a way to express mathematical relationships. You use functions in many real-life situations. Suppose you go to the store to purchase compact discs (CDs). The amount of sales tax you pay *depends* on the amount of money you spend for the CDs. Since the amount of sales tax *depends* on the total cost, you can say sales tax is a function of the total cost.

If you go on a trip, then the distance traveled *depends* on the speed of the car and the time spent traveling. If you travel at a constant speed, then the distance traveled is a function of time alone. When you get a job, you may be paid by the hour. Then your earnings will be a function of the number of hours worked. This textbook will help you use functions to explore and solve problems arising in real-life situations.

Top Ten Reasons To Study Functions

10. You need this course.

9. Functions are in the textbook.

8. Your mother wants you to.

7. Functions will be on the final exam.

6. You need to do something with your graphing calculator.

5. You want to impress your friends.

4. There is a direct correlation between salaries and the amount of mathematics studied.

3. Functions will help you solve problems encountered in your major.

2. Functions can be used to explore real-world phenomena.

1. Functions help you make better decisions both on the job and in everyday life.

The formulas that express the area of a square and circumference of a circle are two functions we encounter frequently. We will explore these in the next two investigations.

The Area of a Square

Equipment needed:
    One-centimeter graph paper on page 28
    Straight edge

Draw squares with sides of 2, 4, 6, 8, and 10 centimeters on a sheet of one-centimeter graph paper. Count the number of square centimeter units in each of the squares drawn. Record your data in the following table:

Length of Side of Square	Area = Number of Square Centimeter Units
2	
4	
6	
8	
10	
s	?

What is the relationship between the area of the square and the length of its side?

We say that the area, A, of a square is a *function* of the length of its side, s. We write this as $A(s)$. This does not mean to multiply A times s. This is an example of functional notation and is read, "A of s." Write a formula for this function based on your data.

$$\text{Area} = A(s) =$$

The Circumference of a Circle

Equipment needed:
    Lids
    String
    Ruler

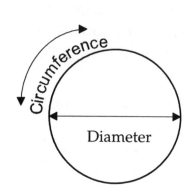

Carefully measure the diameter and circumference of each lid and record the measurements in the table below.

Diameter	Circumference	C + D	C - D	C x D	C/D

Once you have measured the circumference and diameter of the lids, complete the table by finding the sum, difference, product, and quotient of each pair of numbers letting C represent Circumference and d represent the length of the diameter. What patterns do you see?

We say that the circumference of a circle is a function of the length of its diameter. We write this as $C(d)$. As before this doesn't mean to multiply C times d, but is another example of functional notation and is read as "C of d."

Write a formula for this function based on your data.

$$\text{Circumference} = C(d) =$$

Applications Of Functions

Functions can be applied to problems in economics, physics, chemistry, biology, sociology, and numerous other areas. Often authors use functions expressed graphically in popular magazines and newspapers to communicate information quickly and concisely.

This graph dramatically demonstrates the total number of fatal crashes per 100 million miles as a function of the drivers' age. Source: "Safe Driving: The rookies and the veterans," *USAA Magazine*. April 1995.

Source: Insurance Institute for Highway Safety
Reprinted from *USAA Magazine*, April 1995, USAA, San Antonio, TX © 1995.

Often, newspapers hide functions in the paragraphs of an article. It may be more difficult to find a function described in words. This paragraph describes the number of births to unwed mothers as a function of year. Source: *Messenger- Inquirer*. Owensboro, KY. June 10, 1995.

Murphy Brown isn't alone

...Overall, the National Center for Health Statistics said, the unmarried birth rate rose 54 percent from 1980 to 1992. There were 29.4 births per 1,000 unmarried women aged 15-44 in 1980, and 45.2 births per 1,000 in both 1991 and 1992. ...
© 1995 Owensboro Messenger-Inquirer. Reprinted with permission.

The graph below demonstrates the cigarette consumption per U.S. adult per year as a function of the year. Source: *Scientific American.* May 1995.

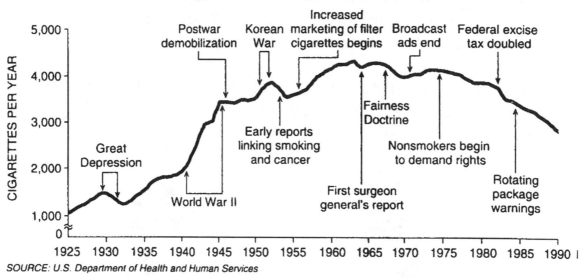

Cigarette Consumption per U.S. Adult

From "The Global Tobacco Epidemic," by Carl E. Bartecchi, Thomas D. MacKenzie and Robert W. Schrier. Copyright © 1995 by Scientific American, Inc. All rights reserved.

Each morning investors can examine graphically the Dow Jones Industrial average from the previous business day as a function of time. Source: "Moneyline," *USA Today.* Thursday, January 25, 1996 .

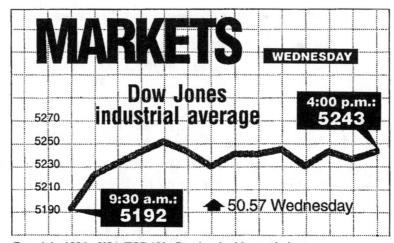

Often more than one function is graphed on the same axes to allow the reader to compare and contrast different situations. This illustration shows depletion curves with production as a function of time for three different assumptions. Source: Miller, *Living in the Environment.* CA: Wadsworth. 1992.

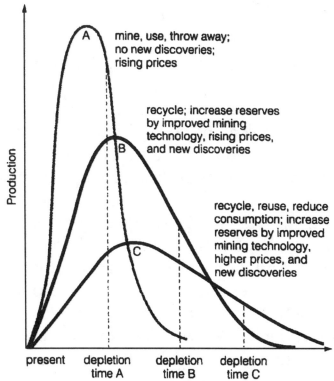

Reprinted with permission of Wadsworth Publishing Co. from *Living in the Environment* by Miller, © 1992.

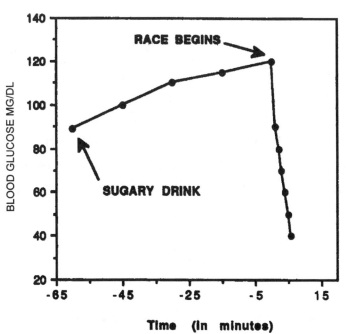

To the left is a graph that shows what occurs in the blood when a runner ingests a sugary drink about an hour before a race. Blood glucose concentration is a function of time. Source: Evans and Rosenberg, *Biomarkers: The 10 Keys to Prolonging Vitality.* NY: Simon & Schuster. 1991.

Reprinted with the permission of Simon & Schuster from BIOMARKERS by William Evans, Ph.D., and Irwin H. Rosenberg, M.D. Copyright © 1991 by Dr. Irwin H. Rosenberg, Dr. William J. Evans, and Jacqueline Thompson.

The illustrations on this page and the previous pages are just a few examples of functions applied to real-world situations. Later you will be asked to find some examples in your field of interest.

Representation of Functions

Functions can be described in various ways — verbally, algebraically, numerically, and graphically. For instance, verbally you could discuss or write about the function that represents the area of a square. Numerically, you could look at the table of data collected from the experiment. Algebraically, you could give the formula for the area, $A(s) = s^2$. Graphically, you could sketch a graph of the function.

Since x and y are used so frequently in mathematics, the developers of graphing calculators use only x and y as variables for graphing. However, you may find it easier to use variables that remind you of the original meaning. For example, we often use A to represent area. In fact, a variable does not have to be just one letter. We could use the word "area" as a single variable. We will often translate our original function such as, $A(s) = s^2$ into $y = x^2$ which is the way to use the calculator.

Both $A(s) = s^2$ and $y = x^2$ indicate you should associate a number with its square.

Ordered pairs of the form (x, y) can be generated from a function of the form $y = f(x)$ by selecting values of x and associating them with their corresponding $y = f(x)$. Note that order is important; (c, d) is not the same ordered pair as (d, c). The ordered pairs $(-2,4)$, $(-1,1)$, $(0,0)$, $(1.1,1.21)$, $(2.5,6.25)$, and $(3,9)$ are some elements of the function where the first member of the ordered pair is a number and the second member is the square of the number. The following table also shows these ordered pairs.

x	-2	-1	0	1.1	2.5	3
$y = x^2$	4	1	0	1.21	6.25	9

The first three pairs of values from the table could also be written in functional notation as $f(-2) = 4$, $f(-1) = 1$, and $f(0) = 0$ if f is the name assigned this function.

✎ Write the last three pairs of values from the table using functional notation.

The figure shows the same pairs of values displayed using the STAT feature on the TI-82. Later you will have the opportunity to use the STAT feature of your calculator to enter and plot points.

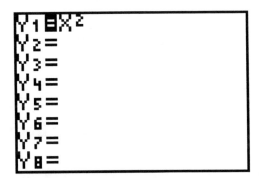

L₁	L₂	L₃
-2	4	▬▬▬▬
-1	1	
0	0	
1.1	1.21	
2.5	6.25	
3	9	

L₃(1)=

Algebraically we write this function as $y = x^2$. On the TI-82, we enter this function using Y= and X,T,Θ. The following display shows the calculator representation of the function.

Y₁⊟X²
Y₂=
Y₃=
Y₄=
Y₅=
Y₆=
Y₇=
Y₈=

Here are two graphs of the function $y = x^2$. The graph on the left was drawn using a computer program while the one on the right was drawn using the TI-82. Drawn on the same scale, the differences in these graphs are due to the finer resolution of the computer screen compared to your TI-82 screen.

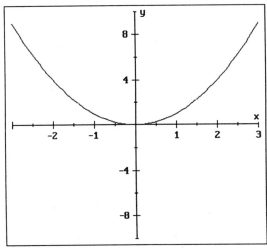

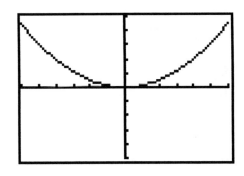

Evaluating Functions Numerically

To use your graphing calculator effectively to evaluate functions, you need to understand the order of operations, both as you would perform the operations and how the calculator interprets your instructions.

When performing operations on algebraic expressions, you should use the order of operations learned in earlier mathematics courses.

Order of Operations

- Perform any operations within the *parentheses*.

- Simplify any numerical expressions involving *exponents*.

- Do any *multiplication* and *division* as they occur from left to right.

- Do any *addition* and *subtraction* as they occur from left to right.

You Try It

Simplify the following expressions without using your calculator.

1. $4 - 2 + 3$

2. $3 + (4 + 2) \div 3$

3. $\dfrac{6}{6}(5)$

4. $\dfrac{3}{4} \cdot \left(\dfrac{4}{9}\right)^2 + \dfrac{1}{2}$

Calculations on the TI-82 are performed on what is called the home screen. If the TI-82 shows a different screen, you should press 2nd MODE, which represents QUIT, to return to the home screen. If you have calculations that you no longer need, you can press [CLEAR] to erase all the calculations from the home screen. The location of these keys is shaded on the TI-82 to the right.

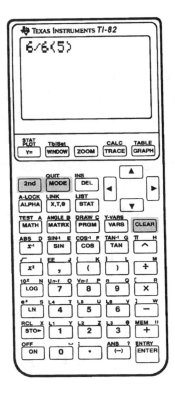

The TI-82 Graphics Calculator uses an order of operations that is very similar. However, try entering $\frac{6}{6}(5)$ as $6 \div 6(5)$. What happened? Did you get the same answer? The TI-82 did not interpret $6 \div 6$ as $\frac{6}{6}$, but calculated 6 times 5 and then divided 6 by 30. You will avoid this situation if you **always use parentheses around any fraction**.

✎ Use your calculator to perform the same calculations. Did your answers match?

You Try It

Using your graphing calculator, evaluate each of the following expressions.

1. $\dfrac{11}{12} - \dfrac{7}{8}$

2. $\dfrac{3}{4} \cdot \left(\dfrac{11}{12} - \dfrac{7}{8} \right) + \dfrac{5}{16}$

3. $-3^2 + (-3)^2$

4. $\dfrac{2+3}{6+4}$

5. $\dfrac{4}{3}\pi$ Check your calculator answers by estimating.

We will use a graphing calculator to examine functions both numerically and graphically. In later units, we will make extensive use of the graphing features of the calculator. It is important that you use your graphing calculator efficiently and appropriately. For example, your supervisor at work will probably not be amused if you take out your TI-82 to calculate 5 times 30. Throughout the book you should ask yourself, can I determine this answer more easily and quickly with or without my calculator?

We will use functional notation when evaluating functions at given values. For example, if $f(x) = x^2$, $f(-2) = (-2)^2 = 4$. What is $f(3)$?

You Try It

1.   If $f(x) = 2x - 3$, what is $f(-2), f(2),$ and $f(0)$?

2.   If $g(x) = 4 - 4x$, what is $g(0), g\left(\frac{1}{2}\right),$ and $g(-4)$?

3.   If $\beta(r) = \frac{2}{3}r$, what is $\beta(6), \beta(0),$ and $\beta(-12)$?

Earlier you found that the area of a square was given by the function

$$A(s) = s^2$$

Now use the calculator to find the area of squares with sides of 14 inches; 2378 centimeters; and 0.00098 miles.

After evaluating the functions, does your screen look similar to the one at the right?

```
14²
               196
2378²
          5654884
0.00098²
        9.604E-7
```

Scientific notation allows us to handle very large and very small numbers. The last entry on the calculator screen shows 9.604E -7. This illustrates the calculator version of scientific notation. The E -7 means to multiply the 9.604 by 10^{-7}. In mathematics and science, we write this as 9.604×10^{-7}. You could also write this number as 0.0000009604.

Numerically, the answers on your calculator are accurate, but are not complete for real-world applications. These are not the answers you should record. The answers for real-world applications should always have units. The complete answers are 196 square inches, 5,654,884 square centimeters, and 9.604×10^{-7} square miles, respectively.

You could write these results as:

$A(14 \text{ inches}) = 196 \text{ inches}^2$

$A(2378 \text{ cm}) = 5,654,884 \text{ square centimeters}$

$A(0.00098 \text{ mile}) = 9.604 \times 10^{-7} \text{ square miles}$

You Try It

1. Calculate the area for squares with sides of 33.56 km, 179.44 yards, 0.000987 cm, and 12,900,000 inches.

2. The circumference of a circle is given by $C(d) = \pi d$. Evaluate the circumference for circles with diameters of 3 inches, 4 meters, 157 yards, and 0.0000074 kilometers.

3. Suppose the distance (in feet) traveled by an object after t seconds is given by the function $d(t) = 40000t - 16t^2$. [Note that the constants in this problem have units that would combine with the time units to give answers in feet.] Find the distance traveled after 10 seconds, 100 seconds, and 610 seconds.

4. If you know the circumference of a circle, the area is given by the function $a(C) = \dfrac{C^2}{4\pi}$. Find the area of circles with circumferences of 16 inches, 300 cm, 6π feet.

    What is $a(2\pi r)$?

Radicals and Exponents

Functions often involve exponents and radicals. You studied integer exponents in earlier courses, but applications frequently also involve negative and fractional exponents. We can use the calculator to evaluate exponents and radicals. Complete the following to improve your understanding of radicals and exponents.

Complete the following using your graphing calculator where appropriate.

1. $2^{-1} =$ _____

2. $\left(\dfrac{1}{2}\right)^{-1} =$ _____

3. $2^{-2} =$ _____

4. $\left(\dfrac{1}{2}\right)^{-2} =$ _____

5. $3^{-1} =$ _____

6. $\left(\dfrac{1}{3}\right)^{-1} =$ _____

7. Complete the following: If a is a positive real number and r is a rational number, then $a^{-r} =$ _____.

8. a) $\sqrt{25} =$ ____

    b) $25^{\frac{1}{2}} = (25)^\wedge(1/2) =$ ____

9. a) $\sqrt{100} =$ ____

    b) $100^{\frac{1}{2}} = 100^\wedge(1/2) =$ ____

10. a) $\sqrt{36+3} \approx$ ____

    b) $(36+3)^{\frac{1}{2}} = 39^\wedge(1/2) \approx$ ____

11. a) $\dfrac{1}{\sqrt{25+9}} \approx$ _____

    b) $(25+9)^{\frac{-1}{2}} = (34)^\wedge(-1/2) \approx$ ____

12. Complete the following based on the examples above: To compute the square root of a number, I can either use the square root key or I can _____.

13. $\sqrt[3]{8} = 8\hat{}(1/3) =$_____ Is your answer really the cube root of 8? Check your answer by cubing it.

14. $\sqrt[4]{25} \approx$_____ How can you check your answer?

15. Complete the following based on the examples above: To compute any root, I _____.

16. Compare and contrast the following problems and the results.
    $100\hat{}(1/3) \approx$_____ $100\hat{}(0.33) \approx$_____

17. Compare and contrast the following problems and the results.
    $64\hat{}(1/2) =$_____ $8^2 =$_____

18. Compare and contrast:
    $42\hat{}(-1/2) \approx$_____ $\dfrac{1}{\sqrt{42}} \approx$_____

Circumference of an Ellipse

Equipment needed:
   Ruler
   String

You can approximate the circumference of an

ellipse using the formula, $\rho = 2\pi \sqrt{\frac{1}{2}(a^2 + b^2)}$.

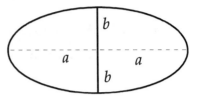

1.   Measure to determine the value of both a
     and b for the ellipse below.

2.   Measure the circumference of the ellipse.

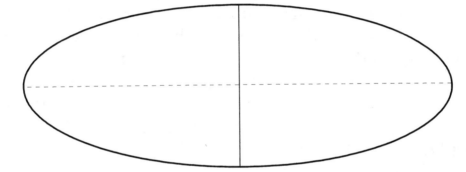

3.   Use the formula to calculate the approximate circumference of this
     ellipse. How does your calculation compare with your
     measurement?

4.   Calculate the approximate circumference of an ellipse with $a = 2.5$
     meters and $b = 1.5$ meters.

You Try It

1. The radius of a sphere with volume V can be determined using the function, $r(V) = \left(\dfrac{3V}{4\pi}\right)^{\frac{1}{3}}$. What is the approximate radius of an orange with a volume of 270 cubic centimeters?

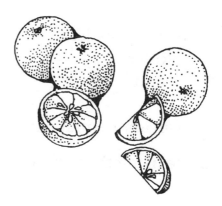

2. Scientists use carbon-14 dating to determine the age of fossils. Live plants continually replenish carbon from the atmosphere. When the plants die, the accumulated carbon-14 begins to decay. If the accumulated carbon-14 is an amount, A, then t years after it died the amount of carbon-14 remaining is given by the formula $C = A \cdot 2^{-t/5600}$. How much carbon-14 will remain after 100 years if the initial amount was 10 micrograms? After 1000 years? After 10,000 years? After 1×10^6 years?

3. The body surface area in square meters for a baby who is 50 centimeters long and weighs w kilograms is given by $B(w) = \sqrt{\dfrac{50 \text{ cm} \cdot w \text{ kg}}{3600 \frac{\text{cm·kg}}{\text{m}^4}}}$.

    What is the body surface area to the nearest hundredth of a square meter of a baby who weighs 3.5 kg?

Summary

You now should agree that functions are useful in expressing mathematical relationships. You have also seen that you can derive functions from experiments and use functions to describe connections in the real world.

You have seen examples of four ways to present functions:
- algebraically;
- graphically;
- numerically; and
- verbally.

One of the goals of this book is to present the tools to use as you explore and solve problems arising from real-world situations. Your graphing calculator will be one of these tools.

If you remember the essentials below, it will make your later work easier.

1. Functions can be written using functional notation, such as $f(x) = 2x^2$. This function would be read as "f of x equals two x squared."

2. Circumference of a circle as a function of diameter: $C(d) = \pi d$

3. Area of a Square as a function of the length of one side: $A(s) = s^2$

4. Complete answers to real-world problems should include units.

FRANK AND ERNEST

FRANK & ERNEST reprinted by permission of Newspaper Enterprise Association, Inc.

Problems for Practice

1. Find the definition of a function in two different dictionaries. Copy both definitions, then state which you think is the more mathematical. Identify both sources.

2. Find at least three examples of functions in books, magazines, or newspapers. Bring copies of the examples to class. Identify each source.

3. Interview professors in your major area (or your favorite professor) and ask for examples of how their discipline uses functions. Bring copies of the examples to class. Identify each source.

4. The volume of a sphere as a function of its radius is $V(r) = \frac{4}{3}\pi r^3$. Find the volumes in cubic units of spheres with radii of 4 in, 15 cm, 0.00235 km, and 125,000,000 miles.

5. The length of the hypotenuse of a right triangle with one side of 4 inches is given by the function $H(x) = \sqrt{16 + x^2}$ where x is the other side of the triangle in inches. Find the hypotenuse for triangles if the other side is 4 inches, 44 inches, or 444 inches.

6. To convert a temperature in Celsius to Fahrenheit, use the formula $f = \frac{9}{5}c + 32$. You could say "multiply the Celsius temperature by $\frac{9^\circ F}{5^\circ C}$ and then add 32° F." Find the Fahrenheit temperature when the Celsius temperature is −20° C.

7.   The average salary for a public school teacher in the United States during the years from 1980 through 1991 can be closely approximately by the function, $S(t) = 1526.6t - 3006594.9$ where t is the year. If this trend continues, what should the average salary of a public school teacher be in 2001? Does your answer seem reasonable? Why or why not?

8.   To measure the rate calcium is absorbed by the body, a subject is injected with a certain amount of a radioactive isotope of calcium. After t days, the concentration of radioactive calcium remaining is $C(t) = t^{\frac{-3}{2}}$ mg/dl. What is the concentration of radioactive calcium in the body after 5.2 days?

9.   The surface area of a sphere in square units is a function of its radius, and can be represented by $S(r) = 4\pi r^2$.

   a)   Find the surface area of a sphere with a diameter of 12.25 inches.

   b)   The moon is almost spherical in shape. Find the surface area of the moon if its radius is 1738 kilometers.

10. Calculate (round to the nearest hundredth):

   a) $\dfrac{5}{\sqrt{8}}$

   b) $\sqrt[4]{81}$

   c) $\dfrac{\sqrt[3]{25}}{4}$

11. It is reported that the maximum speed, v in miles per hour, a car can travel around a curve of radius r feet without skidding is given by the function $v(r) = \sqrt{2.5r}$. What is the maximum speed a car can travel around a curve of radius 250 feet? Approximately, what radius would be needed for a car traveling 65 miles per hour? What decimal fraction of a mile does your answer represent? Is it reasonable to expect highway engineers to construct curves of that radius?

12. The volume of a cube is given by the function $V(s) = s^3$ where s is the length of one side. Use functional notation to indicate the volume of cubes with side lengths of 3 inches, 33 inches, and 333 inches.

If the volume of the cube is 32,768 cubic centimeters, what is the length of one side?

13. Health care workers who need to know the body surface area in square meters for a man if he is 68 inches tall and weighs w pounds use the function $B(w) = \sqrt{\dfrac{68 \text{ in} \cdot w \text{ lb}}{3100 \frac{\text{in} \cdot \text{lb}}{\text{m}^4}}}$. What is the body surface area to the nearest hundredth of a square meter of a man who weighs 170 pounds?

⌘14. The distance $d(h)$ a person can walk in h hours walking at 2 miles per hour is a function. Complete the table below.

h	0.5	1	2	3	3.5	5
$d(h)$						

Write a formula for this function based on your data.

$d(h) =$

⌘15. Perimeter of a square

Equipment needed:
    Ruler

a.   Measure the length of one side of each of the squares drawn below. Record your data in the table on the next page.

b.   Measure the perimeter of each square. Record your data in the table.

Square
A

Square B

Square C

Square D

Square E

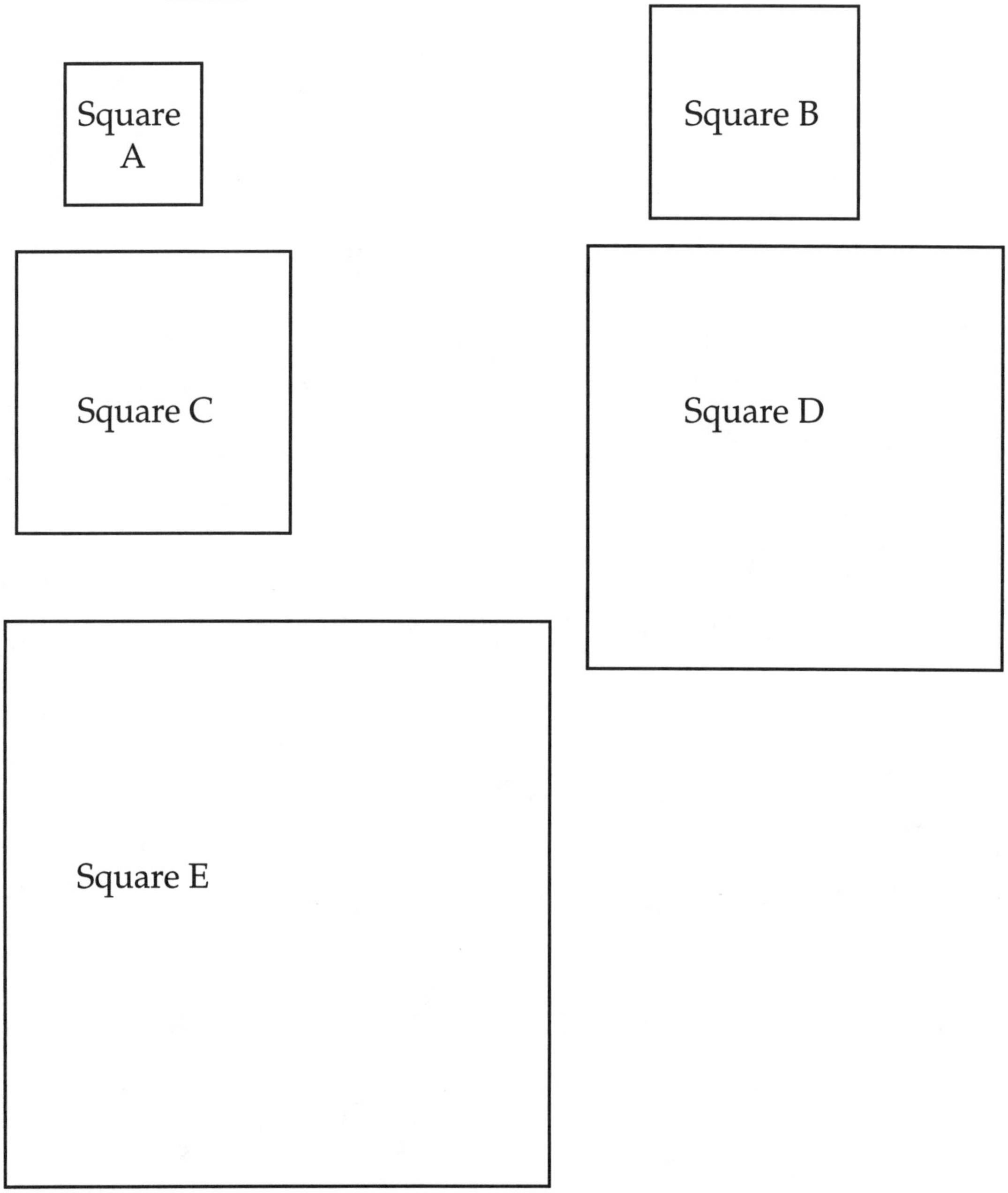

Square	Length of Side of Square	Perimeter
A		
B		
C		
D		
E		

What is the relationship between the perimeter of the square and the length of its side?

The perimeter of a square is a function of the length of its side. We write the function as Perimeter = P(s). Write a formula for this function based on your data. The formula will allow you to evaluate the perimeter of a square with a side of any length.

Perimeter = P(s) =

⌘16. Area of a circle

Equipment needed:
    One-centimeter graph paper
    Lids

Draw circles on one-centimeter graph paper by tracing lids of four different sizes. Measure the diameter of each lid and divide by two to get the radius. *Estimate* the number of squares enclosed by each circle and record your data in the following table.

Radius	(Radius)^2	Area = # of squares

What is the relationship between the area of a circle and the length of its radius?

The area of a circle is a function of the length of its radius. We write this as Area = *A(r)*. Write a formula for this function based on your data.

$$\text{Area} = A(r) =$$

One-centimeter graph paper

One-centimeter graph paper

2

Linear Functions

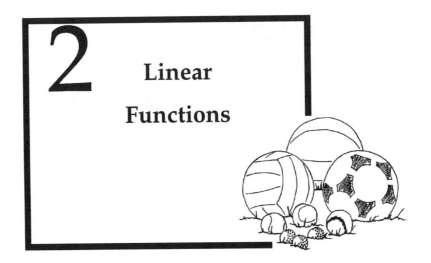

Upon successful completion of this unit you should be able to:

1.   Continue a linear pattern from data in a table;

2.   Determine whether data in a table represents a linear function;

3.   Describe at least three linear functions from everyday life;

4.   Determine the slope and equation of a line;

5.   Graph linear functions; and

6.   Solve application problems involving linear functions including direct variation.

Functions are important in many applications of mathematics. We frequently solve real-world problems using linear functions whose graphs are lines. The skills developed in this unit will be useful when studying other topics.

You will examine linear functions in tables, equations, and graphs. Some activities will involve experiments to collect data; others will involve data given in an equation or in words.

The Bouncing Ball

Equipment needed:
      At least one ball per group
      Meter sticks or tape measures
      Graph paper

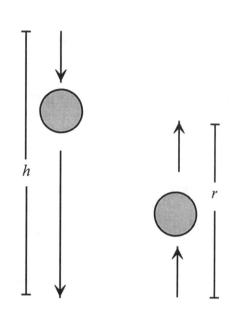

Hold a ball 80 centimeters above the floor, and record this initial height as h. Drop the ball to the floor, and record the height the ball rebounds on its first bounce as R. (You may want to repeat this 2 or 3 times and find the mean, or average, rebound height to obtain a more accurate measure.) Repeat these steps for at least six different drop heights. Record your measurements in the table below, then plot the points on the graph paper.

Ball 1

h						
R						

Now, complete the following table using a different ball.

Ball 2

h						
R						

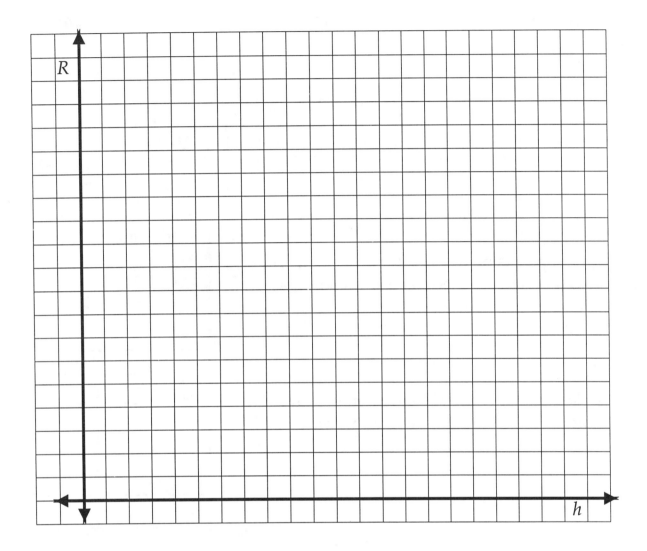

Next we will plot the values on the TI-82 Graphics Calculator. It is important to note that there are usually several different ways to accomplish anything on the TI-82. Read your manual to learn methods other than the ones presented.

Plotting Points on the TI-82

Before you plot your data, plot the data in the following table (data from another class):

h	80	100	70	60	50	40
R	54	74	47	40	24	18

To plot the points on the TI-82 graphics calculator, press STAT to display the screen shown here.

```
EDIT CALC
1:Edit…
2:SortA(
3:SortD(
4:ClrList
```

Press ① or ENTER to edit the entries in the List. Your calculator screen should look like this one.

```
L1    L2    L3
█████ ----- -----



L1(1)=
```

If there are entries in the List, press ▲ until the L_1 is highlighted. Then CLEAR ENTER to erase all entries in the first list. Press ▶ ▲ CLEAR ENTER to erase L2. Press ◀ until $L_1(1)$ = shows at the bottom of the screen.

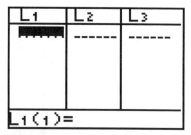

```
L1    L2    L3
70.5  42    1.6786
66    40    1.65
64    37.5  1.7067
66    39    1.6923
60    37    1.6216
69    42    1.6429
69    39    1.7692
L1={70.5,66,64,…
```

Once there are no entries in L_1 and L_2, type the first h-value, 80. Press ENTER. Continue this process until you have entered the last h-value.

```
L1    L2    L3
80    ----- -----
100
70
60
50
40
█████
L1(7)=
```

Press the right arrow, ▷. This will move the cursor to the top of the next list, L2. Now enter the R-values in the same order. The TI-82 screen should look similar to this one.

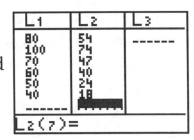

There must be the same number of entries in L1 as there are in L2 if you plan to plot the data as points. Now press Y=. The screen should look like this one.

If not, press CLEAR to erase the first equation. Press ▽, then CLEAR until all equations are erased.

Press 2nd Y= to display STAT PLOTS. Plot 1, 2, and 3 should all display Off at this time. If any plots display On, press 4 ENTER. This will return you to the HOME screen so you will need to press 2nd Y= to display STAT PLOTS again.

Press 1. Although the Off is highlighted, the cursor should be flashing at the On. If so, press ENTER to turn Plot 1 on. The Type: should be ⌐ (Scatterplot). For this plot, the Xlist indicates where the h-values are located (L1). The Ylist indicates where the R-values are located (L2). Choose any of the three marks, or use the one your instructor suggests.

If the type is not ⌐ , arrow down and press ENTER to select it. Once your TI-82 screen matches the screen on the left, press WINDOW to define the portion of the coordinate system that is displayed on the screen.

To plot the given ordered pairs on graph paper you would consider the largest and smallest values for the axes. The same is true for the graphing calculator. What is the smallest *h*-value? What is the largest *h*-value?

What is the smallest *R*-value? What is the largest? Remember that *x* will represent *h* and *y* will represent *R* on the TI-82.

Let's plot the points on the portion of the coordinate system between –5 and 110 on the horizontal axis and –5 and 80 on the vertical axis. Starting a little less than 0 allows the axes to be displayed. So we will use –5 for the Xmin value and 110 for the Xmax value in the viewing window. Also, we will use –5 for the Ymin and 80 for the Ymax values.

Although the scale is not critical at this time, we may want to change the scale to 10 for both the Xscl and the Yscl.

Press ▼(-)5 ENTER to change the Xmin. Press 1 1 0 ENTER for Xmax. Continue this process until the screen matches the screen displayed. Be careful and press the negative, (-), and not the minus, —, when entering these values.

```
WINDOW FORMAT
Xmin=-5
Xmax=110
Xscl=10
Ymin=-5
Ymax=80
Yscl=10
```

Now press GRAPH to see the plot.

Before you plot your ball drop data, clear L₁ and L₂. Press STAT ENTER. To clear the entries in the list, press ▲ until the L₁ is highlighted. Then CLEAR ENTER to erase all entries in the first list. Press ▶ ▲ CLEAR ENTER to erase L₂. Enter the data as before, set the window, and plot the points.

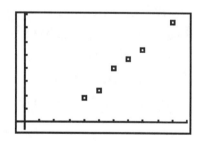

```
L1      L2      L3
80      54      ------
100     74
70      47
60      40
50      24
40      18
------  ------
L1=(80,100,70,6...
```

Often during this course, you will plot points to determine the shape of a graph.

✎What shape does the graph of your data and the graph above suggest?

The relationship between initial height and rebound height can be approximated by a straight line. We can say the rebound height *depends* on the initial height from which the ball was dropped.

Another common linear relationship is distance versus time. The relationship between the distance traveled at a constant rate and the amount of time traveled is a linear function. We can say the distance traveled *depends* on the time spent traveling.

Distance vs. Time

Dr. Warren called to say he was running late. He set his cruise control at 45 miles per hour to avoid getting stopped for speeding.

How far would he go in zero hours? In one hour? In two hours? In 3 hours?

Construct a table where *t* represents time and *d* represents distance traveled.

t	0	1	2	3			
d							

Notice as the time increases by 1 hour, the distance traveled increases by 45 miles. Add the distance traveled for 4 hours, 5 hours, and 6 hours to your table.

Plot these points on a coordinate system using the vertical axis for distance and the horizontal axis for time.

> If you read, "draw a graph of D vs. T" remember D is the label for vertical axis and T is the label for the horizontal axis.

Using a straight edge, draw a line through the points. This is a linear function.

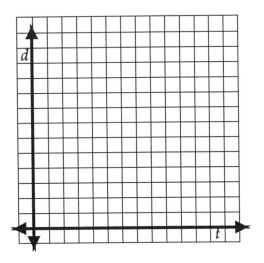

✎ Using your graph, estimate how far he can travel in one hour and 30 minutes. _____

In 2 hours and 45 minutes? _____

How far can he go in 15 minutes? _____

About how long will it take him to travel 100 miles. _____

Are you confident your answers are reasonable? Why or why not?

You Try It

If Dr. Warren traveled at a constant speed of 60 miles per hour, how far would he travel in 0 hours, $\frac{1}{2}$ hour, 1 hour, and $\frac{3}{2}$ hours?

t							
d							

1. As the time increases by _____ hours, the distance traveled increases by _____ miles. Complete the table for 2 hours, $2\frac{1}{2}$ hours, and 3 hours.

---

2.   Plot the points on the graph using the same scale as before. Connect the points with a line.

3.   As the time increased by 1 hour, by how many miles does the distance traveled increase? Compare and contrast your answer to this question and your graphs.

4.   Repeat this process for a constant speed of 30 miles per hour.

t							
d							

5.   How does the graph change as the speed increases?

6.   Plot the points for each speed on the TI-82. Using a different mark for each speed you might display a graph like this.

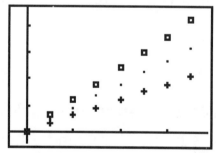

If the points are connected smoothly, each graph is a line. In each table if the difference between the times was constant, the difference between the distances traveled was constant. You can use either the constant differences or a graphed line to predict distance traveled for other times.

The plots above also demonstrate that the line gets steeper as the rate of speed increases.

Each of the following tables represents a linear function. (a) Study the pattern in each of the following tables, and complete each. (b) What is the constant difference for the x-values? (c) What is the constant difference for the y-values? (d) Plot the points. Do the points lie in a line?

1.

Difference in x								
x	−3	−2	−1	0	1	2	3	4
y	−9	−6	−3	0				
Difference in y								

2.

Difference in x								
x	−6	−4	−2	0	2	4	6	8
y	10	8	6	4				
Difference in y								

3.

Difference in x								
x	−3	−2	−1	0	1	2	3	8
y	−0.5	0	0.5	1				
Difference in y								

In any linear function, the dependent variable, y, changes (increases or decreases) at a constant rate as the independent variable, x, changes at a constant rate.

Walking the Line

To help you develop an intuitive understanding of lines and motion, your teacher may use the Calculator-Based Laboratory (CBL) and a motion detector to allow you to walk the line. Further development can be accomplished by completing the following problems.

1. Describe to your friend how to walk so that the walking motion matches the graph.

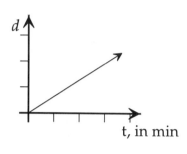

2. What do you do to create a horizontal line on a distance versus time graph?

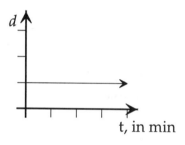

3. How do you walk so that your motion matches the graph? Where do you start?

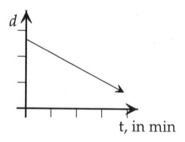

4. Compare and contrast the graph of Pablo's walk (- - - -) with the graph of Jim's walk (_____).

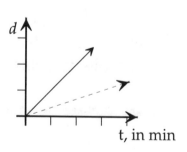

Off To School

Toni and John walk to school in their hometown of Paris, Kentucky. One morning, John walks to the Sports Store, spends a few minutes looking at the tennis shoes and then runs to school so he can shoot a few baskets before school starts. Toni spends a few minutes talking on the phone before she leaves, runs to the Sports Store, where she spends a couple of minutes looking at a tennis racket she likes, and then walks to school.

1)   On a graph of distance versus time, what point could represent the moment when the person is at home at time equal to 0?

2)   How can you represent time spent at home talking on the phone?

3)   Describe the similarities and differences between the lines representing a person running and a person walking.

4)   Which of the graphs below describe John's trip to school?

5)   Which one best describes Toni's?

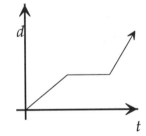
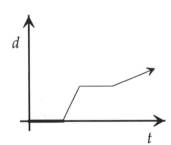

You Try It

Match the following situations with the best possible graph from the ones below. Write a situation for the remaining graph.

1. Chrissy awakens and quickly gets ready for work. She walks downstairs to the garage and drives to work.

2. Bill awakens, walks upstairs to wake his daughters, then downstairs to the kitchen for breakfast. After breakfast he drives them to school.

3. Bobby awakens and dresses to get ready for work. He walks down the hall to wake his children, then walks to the kitchen to sit and read the paper while he drinks his coffee. Then he quickly walks out to his car and drives to work.

4. _____

   _____

   _____

   _____

A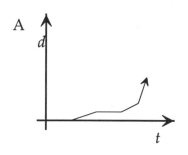

B

C

D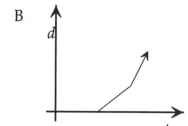

Direct Variation

A linear function, like Dr. Warren's trip, in which pairs of numbers vary directly by a constant rate is also called **direct variation**.

Peter McNary, an entry-level Water Testing Technician, earns \$12/hour. Complete the table.

Hours Worked	0	1	2	3	4	5	6	7	8
Salary Earned									

The salary earned *depends* on the number of hours worked. Consider the rate of pay earned to the number of hours worked (salary/hour):

$$\frac{\$12}{1hr} = \frac{\$24}{2hrs} = \frac{\$36}{3hrs} = \frac{\$48}{4hrs}$$

Notice the rate of pay is always the same — 12 dollars per hour.

Problems involving direct variation use three variables or more to express the situation. One of the letters is usually k and is used for the constant. For the Peter McNary problem, the constant is \$12 per hour. Note that k is the constant for a situation, but a different rate of pay will have a different k. The other variables represent the information given in the table.

We can write an equation that expresses Peter McNary's salary:

Salary earned $= $ hourly wage $\times$ hours worked,

$$S = kh$$
$$S = 12h$$

In functional notation we write Peter's case as $S(h) = 12h$.

✎ Sketch the graph of the salary vs. the number of hours worked. Label the axes. Notice that the hourly wage indicates the steepness or slope of the line.

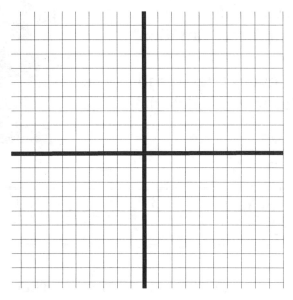

✎ Predict how the line will change if Peter's rate of pay in dollars per hour increases?

Plot the points using STAT PLOT. A quick way to set the WINDOW and plot the points is to press [ZOOM] [9]. [ZOOM] [9] sets the WINDOW so that all points are visible on the screen, although you may want to change the WINDOW later.

You will graph the function $S(h) = 12h$ on the TI-82 using the generic equation, $y = 12x$. The x on the TI-82 always represents the independent variable. The y represents the dependent variable. Because Salary *depends* on the number of hours worked, Salary is the dependent variable, y.

> The x on the TI-82 always represents the independent variable. The y represents the dependent variable.

To use the TI-82 to draw the graph, press ⌨Y= to display the Y= screen. Press 1 2 X,T,θ. Your display should match this screen.

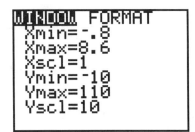

Press GRAPH. Does the line pass through each of your plotted points?

Press WINDOW and record the TI-82 WINDOW using this format: [Xmin, Xmax]$_x$ by [Ymin, Ymax]$_y$. You would record the WINDOW shown here as $[-.8, 8.6]_x$ by $[-10, 110]_y$.

✎ Record your WINDOW: [___ , ___]$_x$ by [___ , ___]$_y$.

> A graph in a window of $[a, b]_x$ by $[c, d]_y$ means the minimum value of x is a, the maximum value of x is b, the minimum value of y is c, and the maximum value of y is d. The choice of scales should be appropriate to the minimum and maximum values stated.

Change Xmax to 12 and Ymax to 150 to see Peter's salary when he works 12 hours.

Press TRACE. In the upper right-hand corner you can see P1. Pressing the right or left-hand arrow (◀ or ▶) causes the cursor to move from plotted point to point. The x and y-values of the point are recorded at the bottom of the screen.

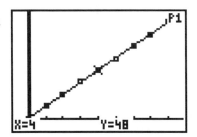

To trace on the function, press the up or down arrow (▲ or ▼). This action moves the tracing cursor from the plot to the function you entered in Y=.

The 1 in the upper right-hand corner indicates the cursor is tracing on the function you entered at Y1=.

As you press ◁ or ▷ the cursor moves from x-value to x-value. At each x-value, the TI-82 calculates the y-value by substituting the x into the function. For example, at $x = 5.6$, the TI-82 evaluates Y₁(5.6) which is 67.2.

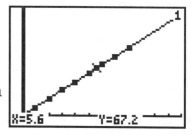

✎ Use the TI-82 to answer the following questions.

1.   What is Peter's salary when he works 12 hours?

2.   Change the WINDOW, then trace to determine Peter's salary when he works 40 hours.

3.   Using the graph, approximate how long Peter must work to make $105? Confirm your answer algebraically.

4.   Graphically, estimate how much money Peter makes working 7.5 hours? Confirm your answer algebraically.

5.   What would happen to the salary if the hours worked are doubled? Halved?

6.   Evaluate, then describe the practical meaning of $S(4)$. $S(52)$.

7.   What would happen to the graph if Peter receives a raise to $14 per hour?

You Try It

Suppose the income tax paid by Eugenios varies directly as the income earned. If Eugenios pays $4 income tax per $100 income, complete the following table.

Income, I	100	200	300	400	500	600	0
Income Tax, T							

1.   Plot the points on your TI-82.

2. $T(I) = 0.04 \cdot I$ is the function that represents this situation. Graph the function on your TI-82.

3.   How much tax will Eugenios pay when he earns $450 a month? 1000 a month?

4.   What happens if his salary is doubled? If your state has an income tax, is this a realistic situation?

Often the problem in direct variation is to find the variation constant, k. If we know an x and the corresponding y, we can determine k if $y = kx$ by substituting x and y, then solving for k.

> If y varies directly as x, then $y = kx$ for some constant k.

Electricity Usage

The utility company measures electricity usage in kilowatt-hours (kwh). The charge to the consumer varies directly as the number of hours of use. If it costs $84 to use 1050 kwh, how much does each kwh cost? What is the charge for 1500 kwh?

The **general equation** that gives the charge, C, as a function of the kilowatt-hours used, h, is $C = kh$.

Find k. Substitute $C = \$84$ and $h = 1050$ kwh into the general equation then solve for k.

$$
\begin{aligned}
C &= kh \\
\$84 &= k \cdot 1050 \text{ kwh} \\
\frac{\$84}{1050 \text{ kwh}} &= k \\
\frac{\$0.08}{\text{kwh}} &= k
\end{aligned}
$$

The **specific equation** is written by substituting this k into the general equation or, for this situation, $C = \dfrac{\$0.08}{\text{kwh}} \cdot h$. The Charge function is written, $C(h) = \dfrac{\$0.08}{\text{kwh}} \cdot h$.

✎ Evaluate $C(1)$ to calculate how much each kwh costs.

✎ Finally, determine the charge for 1500 hours by evaluating $C(1500)$.

You Try It

The coiled spring beneath a park toy supports the weight of a child. The spring compresses 0.8 inches under the weight of a 25-pound child. If the distance (d) compressed varies directly with the weight (W) of the child, how much does the spring compress under the weight of a 65 pound child?

1. What is the general equation?

2. What is the value of k?

3. What is the specific equation?

4. How much does the spring compress under the weight of a 65-pound child?

5. Graph the function, then estimate how much the spring compresses under the weight of a 30-pound child.

6. If the yard toy will not work properly if the spring is compressed more than 3 inches, what is the weight of the heaviest child who should be allowed to use the toy?

After investigating direct variation you may have noticed that there is a relationship between the coefficient of x and the steepness of the line. This relationship is called the **slope**.

The slope of a line is a measure of the line's steepness and is the ratio of the vertical change to the horizontal change or

$$\text{slope} = \frac{\text{vertical change}}{\text{horizontal change}}.$$

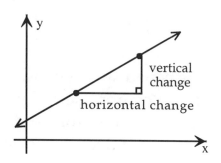

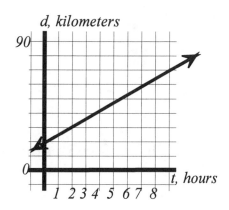

Calculate the slope of this line. First, draw a triangle similar to the one above. For this problem, the slope is the change in distance over the change in time. Therefore, the units will be $\frac{\text{kilometers}}{\text{hour}}$. If t increases by one hour, by how much does d increase or decrease?

You Try It

1. Determine which of the lines below has the greater slope, without calculating the slope. The window is $[-5, 5]_x$ by $[-5, 5]_y$.

a)

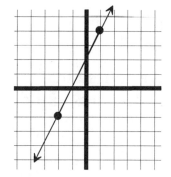

b)

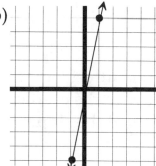

c)

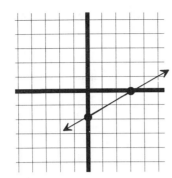

d)

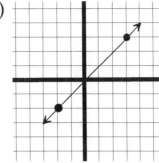

2. Confirm your answer by finding the numerical value of each slope. Label the two points you used to determine the slope with the correct coordinates.

Recall that the slope of a line is the ratio of vertical change to horizontal change. If you have two different points on a line, (x_1, y_1) and (x_2, y_2), you calculate the vertical change by finding the difference in the y-values, $y_2 - y_1$. Calculate the horizontal change by finding the difference in the x-values, $x_2 - x_1$. Use the following formula to find the slope of a line through two points:

$$\text{slope} = m = \frac{y_2 - y_1}{x_2 - x_1}$$

You Try It

Find the slope of the line that passes through each pair of points.
1.   (4, 0) and (5, 8)

2.   (7, –3) and (–3, 8)

3.   (0, 4) and (–1, 9)

4.   (3, 3) and (9, 3)

Next, complete the tables of values for each of the following equations.

5. $y = -2x + 5$

x	−1	0	1	2	3
y					

6. $y = \frac{2}{3}x - 7$

x	−3	0	3	6	9
y					

7. $y = 6x + 13$

x	−3	−2	−1	0	1
y					

Sketch a graph of y versus x for each of these equations.

5.

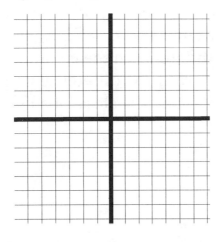

6.

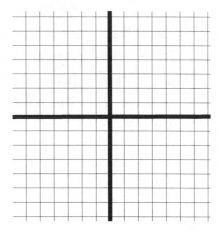

7.

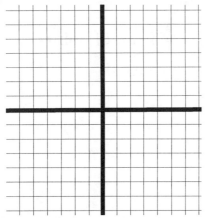

Look back at each table, choose 2 points and calculate the slope for each one. Does the slope change if you choose different points? Can you determine the slope by examining the equation?

Intercepts

The **y-intercept** is the point at which the graph crosses the y-axis.

✎What is the y-intercept for each of the graphs in the **You Try It**?

What is the x-value of each y-intercept?

How can you algebraically determine the y-intercept when you have an equation?

Can you determine the y-intercept by looking at the equation?

The **x-intercept** is the point at which the graph crosses the x-axis.

✎What is the x-intercept for each of the graphs in the **You Try It**?

What is the y-value of each x-intercept?

Can you determine the x-intercept by looking at the equation? If not, can you solve the equation to find the x-intercept?

You Try It

Rebecca is repaying a $400 loan from relatives. The relatives expect no interest and Rebecca can afford to pay them back at the rate of $20 per week. Sketch a graph showing the amount she still owes (A) versus the number of weeks (t) she has been repaying the loan.

1. Complete the table.

Difference in x							
t	0	1	2	3			
A							
Difference in y							

2. Plot the points, then connect, if linear. Label the axes.

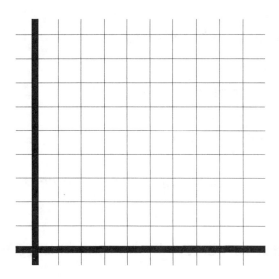

3. What is the slope of the line (include units)? What is the interpretation of slope for this problem?

4. At what point does your graph cross the horizontal axis? _____ This point is the **x-intercept**.

5. At what point does your graph cross the vertical axis? _____ This point is the **y-intercept**.

6. For this problem, what is the significance of the x-intercept and the y-intercept?

7. When will Rebecca owe $200?

8. When will Rebecca pay off the loan?

Since the amount still owed *depends* on the number of weeks, we say that the number of weeks is the independent variable. The amount still owed is the dependent variable. This dependence of one variable on another is one characteristic of a function.

Definition:

A *function* is a relationship between two variables (one dependent variable and one independent variable) where an independent variable value is paired with exactly one dependent variable value.

In any function, for each x there exists one and only one y.

✎ In the "Distance vs. Time" problem, which is the dependent variable and which is the independent variable?

✎ Suppose you go to the store to purchase compact discs (CDs). Let t represent the amount of sales tax you pay, and let C represent the number of dollars spent on the CDs. Which is the independent variable, and which is the dependent variable?

✎ When you get a job, you may be paid by the hour. Then your earnings will be a function of the number of hours you work. Would earnings be represented by the independent or dependent variable?

Let's recap what have you learned to this point.

- How to generate ordered pairs from an equation of a linear function.

- How to determine the slope of a linear function.

- How to graph linear functions.

Now we need to be able to write the equation of linear functions.

The formula for slope can be adapted to allow you to easily write the equation of the line when you know the slope, m, and one of the points, (x_1, y_1).

> The point-slope form of the equation of a line is
> $$y - y_1 = m(x - x_1)$$

Suppose you know that one point is (2, 3) and that the slope is 8. Then you substitute the 2 for the x_1 and the 3 for the y_1. You also know that $m = 8$ so the equation of the line is $y - 3 = 8(x - 2)$. Simplifying, you obtain $y = 8x - 13$.

You Try It

Write the equation of the linear function.

1.   (4, 3) and $m = 5$

2.   (–8, 3) and $m = -1$

3.   (4, 6) and (8, 2)

4.   (5, –1) and (–4, 7)

Return to the Bouncing Ball

Consider the data from the first ball drop on page 30. Does the relationship between initial and rebound height seem to be linear? If so, use the data to write an equation where rebound height depends on initial height. What would be the rebound height if you dropped the ball from 57 centimeters? What height would you have to drop the ball from for the rebound height to be 62 centimeters?

You Try It

Each of the following tables represents a linear function. Study, then complete, the pattern in each.

1.

x	–3	–2	–1	0	1	2	3	4	5	6
y	–10	–7	–4	–1	2					

2.

x	–3	–2	–1	0	1	2	3	4	5	6
y	8	7	6	5	4					

3.

x	–3	–2	–1	0	1	2	3	4	5	6
y	1	3	5	7	9					

4. Write the equation of each of the lines above. Can you use any point you choose that lies on the line? Why?

5. Sketch each of the lines above.

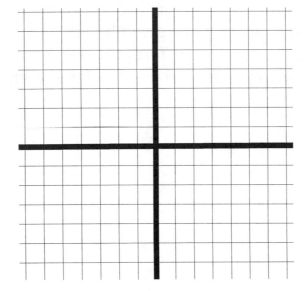

Slope-Intercept Form

For each of the problems in the last *You Try It*, complete the following table:

	Equation	Slope	y-intercept
1.	$y =$ _____	_____	_____
2.	$y =$ _____	_____	_____
3.	$y =$ _____	_____	_____

Consider the information in the table above and make a generalization.

> The slope-intercept form of the equation of a line is given by
>
> $$y = mx + b,$$
>
> where m is the slope and $(0,b)$ is the y-intercept.

You can now find the equation of a line given two points or a point and the value of the slope using the point-slope formula: $y - y_1 = m(x - x_1)$. If you know the slope and y-intercept, you can use the slope-intercept form: $y = mx + b$.

You Try It

Give the slope and y-intercept for the following lines:

1. $y - 8 = 4(x + 3)$

2. $y + 5 = 2(x - 5)$

3. $3y + 7x = 15$

4. $4x - 3y = 24$

The Slope and y-Intercept

Suppose Lynn, a plumber, charges a $50 flat fee plus $37 an hour, for every hour she spends on a job.

1. Identify the independent and dependent variables.

2. If n is the number of hours worked and C is the charge, complete the following table.

n	C
0	
1	
2	
3	
4	
5	
6	
7	
8	

3. Plot these points on your TI-82.

4. What is the rate or slope?

5. What is the charge when Lynn walks in the door (at $t = 0$)?

6. Write an equation for this linear function.

7. Graph the equation on the TI-82.

8. Write the Charge equation in functional notation.

9. What would be the cost equation if she charged a flat fee of $25?
$y = C_2(x) = $_____? What are the slope and the y-intercept?

10. What would be the cost equation if she did not charge a flat fee?
$y = C_3(x) = $_____? What are the slope and the y-intercept?

You Try It

Martin is repaying the $800 he borrowed from his mother to pay for his car insurance. His mother expects no interest, and he can afford to pay back $50 a week.

1. Complete the table.

2. Write an equation.

3. Graph the equation.

4. What will Martin owe after 5 weeks?

5. When will Martin pay off the loan?

Properties of Linear Functions

1. The dependent variable, y, changes (increases or decreases) at a constant rate as the independent variable, x, changes at a constant rate.
2. The points of the graph lie on a line.
3. The function relating the independent and dependent variables can be written $y=mx+b$, where m is the slope and $(0,b)$ is the y-intercept. The slope, m, and b are real numbers.

You Try It

Which of the following represent linear functions? Write the equation of any line.

1.

x	-3	-2	-1	0	1
y	12	7	2	-3	-8

2.

x	5	6	7	8	9
y	-3	-3	-3	-3	-3

3.

x	0	1	2	3	4
y	-2	-1	2	7	14

4. The relationship relating time and fee charged if a baby-sitter charges $3.00 plus $3.00 per hour, where time is the independent variable and fee charged is the dependent variable.

5. The relationship relating distance and time if a car is traveling 40 miles per hour, where time is the independent variable and distance is the dependent variable.

6. The relationship relating rate and time if a car travels 20 miles, where time is the independent variable and rate is the dependent variable.

Vertical and Horizontal Lines

Determine the slope and intercepts of each line below. The window is $[-5, 5]_x$ by $[-5, 5]_y$.

a) 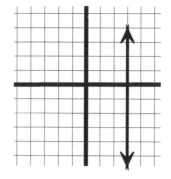   b)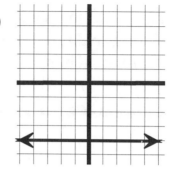

 Do you notice anything unusual, or did you get the answer you expected?

 Try some similar examples, and make a generalization about the slopes of vertical and horizontal lines.

 Write the equation of each line above.

> Vertical lines always have the form, $x = c$, where c is the x-intercept.
>
> Horizontal lines always have the form, $y = b$, where b is the y-intercept.

You Try It

1.   Sketch a graph of the line, $y = 3$.

2.   Sketch a graph of the line, $x = -1$.

 **Summary**

I. Lines

- The slope is a measure of the line's steepness and is the ratio of the vertical change to the horizontal change or slope $= \dfrac{\text{vertical change}}{\text{horizontal change}}$.

- The slope of the line through points (x_1, y_1) and (x_2, y_2) is found using the formula, slope $= m = \dfrac{y_2 - y_1}{x_2 - x_1}$.

- The slope-intercept form of the equation of a line is $y = mx + b$.

- If you know the slope, m, and one of the points, (x_1, y_1), you can write the equation of a line using the point-slope form, $y - y_1 = m(x - x_1)$. The y-intercept, $(0, b)$, could be chosen for the point.

- "Draw a graph of D vs. T" usually means D should be graphed on the vertical axis and T on the horizontal axis.

II. Functions

A *function* is a relationship between two variables (one dependent variable and one independent variable) where an independent variable value is paired with exactly one dependent variable value. This means for each x there exists one and only one y.

III. Properties of Linear Functions

- The dependent variable changes (increases or decreases) at a constant rate as the independent variable changes at a constant rate.

- The points of the graph lie on a straight line.

- The function relating the independent and dependent variables can be written in the form $y=mx+b$, where m is the slope and $(0, b)$ is the y-intercept. The slope, m, and b are real numbers.

ProCats

Memorandum #2   **Lexington, KY 40506**

TO: Team Members

FROM: Your Supervisor

Management is considering selling caps embroidered with the team logo. Each cap costs $2.25 with a setup charge of $150. Please complete the situation analysis and report as outlined below.

The Analysis:

1. Construct a cost function.

2. Graph the cost function and prepare a table of values which might be useful to management.

3. If management plans to sell 500 caps, what will be the actual cost for each cap?

4. What would your team suggest for a retail price? Why?

The Report:

Write a report including the above information and conclusions and submit it to your supervisor. The report should explain your work, your analysis, any assumptions you made, and comments about the reliability of your results as well as your conclusions.

Problems for Practice

⌘1. The relationship between Fahrenheit temperature and the corresponding Celsius temperature is linear. In Celsius, the freezing point of water is 0° C and the boiling point of water is 100° C. In Fahrenheit, these measures are 32° F and 212° F, respectively.

a) Plot the two known facts as Fahrenheit versus Celsius on the graph paper below. Connect the two temperatures with a line.

b) Using the graph, estimate the temperatures below.
  70° C = _____° F   _____° C = –5° F
  13° C = _____° F   _____° C = 98.6° F

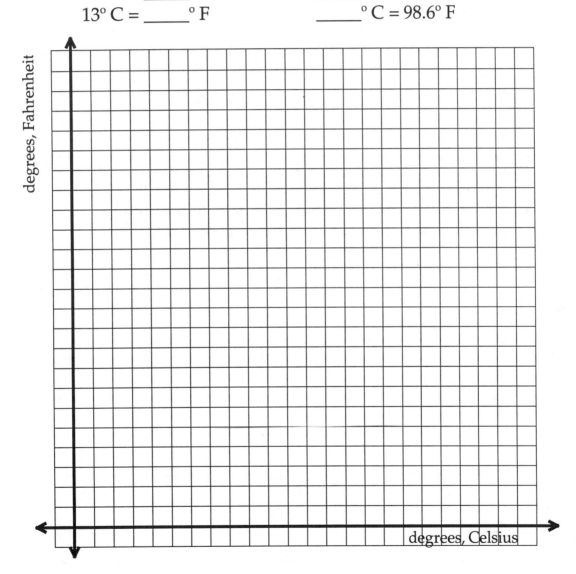

#2. Your maximum heart rate during aerobic exercise is a linear function of your age.

Age in years, a	20	25	30	35	40	45	50	55	60
Heart rate in beats per minute, r	200		190		180				

    a)   Complete the table.

    b)   Plot the points.

    c)   What would be the age of a person with a maximum heart rate of 151?

    d)   Describe the relationship between maximum heart rate and age.

Each of the tables in # 3 – 6 represents a linear function. (a) Study the pattern in each and complete the pattern. (b) What is the constant difference for the x-values? (c) What is the constant difference for the y-values? (d) Plot the points. Do the points lie in a line?

3.

Difference in x							
x	5	6	7	8			
y	13	15	17	19			
Difference in y							

4.

Difference in x							
x	−1	−3	−5				
y	5	1	−3				
Difference in y							

5.

Difference in x							
x	−1	−3	−5				
y	−5	−9	−13				
Difference in y							

6.

Difference in x							
x	−3	0	3				
y	−19	−7	5				
Difference in y							

Match each of the situations in # 7 – 9 with the graph that best corresponds to it. Explain your reasoning for each.

7.   A driver traveled to a certain point, stopped to rest, then continued driving at the same rate.

8.   A driver saw a police cruiser on the side of the road and reduced the speed of the car.

9.   A driver traveled to a certain point, stopped to rest, then continued driving at a slower rate.

(a)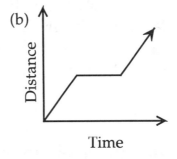

(b)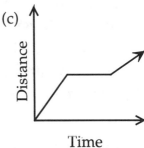

(c)

10.   Sketch a graph that would correspond to the following situation: A driver spotted a restaurant while driving, but missed the exit. He took the next exit, and traveled to the restaurant (in the opposite direction). He stopped at the restaurant to eat, and continued his trip at the same rate.

11. The following are *distance versus time* graphs for two cars traveling at various rates. Write a situation that would fit each graph.

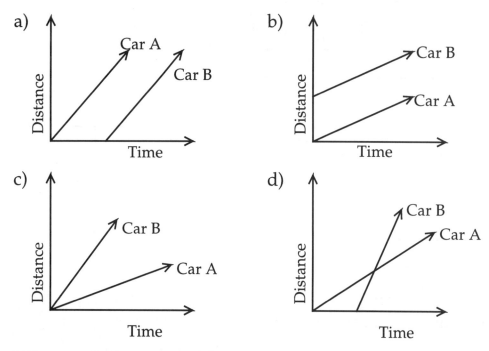

12. What might be reasonable units for the slope or slant of each line above? Notice that the two lines in the last graph intersect — what does this point of intersection represent?

13. Kay leaves Maysville for Paducah traveling at a rate of 55 mph. Dana leaves Maysville for Paducah 1 hour and 30 minutes later, traveling at a rate of 70 mph. Sketch a graph that would represent this situation and find out how long it will take Dana's car to overtake Kay's car.

14. On a 240-mile trip, Beth drove above the speed limit at an average speed of 60 miles per hour for the first 120 miles and at an average speed of 40 miles per hour for the second 120 miles. What was her average speed for the entire trip? Sketch a graph of the situation to help set up the problem.

To help you review elementary algebra, solve the following linear equations for the indicated variable.

15. $2x - 3 = 5$ for x

16. $\frac{4}{3}x - 3 = \frac{7}{5}$ for x

17. $1 - 2p = 4$ for p

18. $2x + 3y = 4$ for y

19. $\frac{1}{2}x - \frac{2}{3}y = 4$ for y

20. $d = rt$ for t

For each of the following problems 21 – 28, (a) state the general equation, (b) determine the constant k, and (c) solve the problem.

21. If V varies directly as I, and V = 24 when I = 3, find V when I = 5.

22. If a varies directly as c, and $a = 7.5$ when $c = 5$, find a when $c = 13$.

23. The volume of a sphere varies directly with the cube of its radius. If a sphere with a radius of 2 ft has a volume of approximately 33.51 cubic ft, approximately what is the volume of a sphere with a radius of 7 ft?

24. Sales tax varies directly as the cost of a taxable item. If the sales tax on a $350 item is $28, find the sales tax on a $4500 item.

25. The surface area of a cube varies directly as the square of an edge. If the surface area is 96 square cm when the edge is 4 cm, find the surface area when the edge is 7 cm long.

26. When an object is dropped, the distance it falls in t seconds varies directly as the square of t. If an object falls 4 feet in 0.5 seconds, how far will it fall in 3.5 seconds?

27. Suppose the pressure on the bottom of a swimming pool varies directly as the depth of the pool. If the pressure is 125 lb/sq. ft. when

the water is 2 feet deep, what is the pressure when the water is 6 feet deep?

28.   The area of an equilateral triangle varies directly as the square of the side. If the area is $9\sqrt{3}$ square inches when the side is 6 inches long, find the approximate area when the side is 4.75 cm long.

✻29.   Commercials tend to use up a fixed portion of the total broadcast time of a television program. For example, if there are 15 minutes of commercial time every hour, then the commercial time is $\frac{15}{60}$ or $\frac{1}{4}$ of the total time. The next time you watch television, determine what portion of the total broadcast is commercial time.

a.   Express this relationship in an equation where commercial time depends on total broadcast time.

b.   Watch television during a different time period – does your equation still hold?

30.   A temperature in Celsius varies directly as the difference between the Fahrenheit reading and 32° F. If a Celsius thermometer reads 37° when the Fahrenheit one reads 98.6°, what will be the temperature in degrees Celsius when the Fahrenheit thermometer reads 5°.

Plot the two points on graph paper in # 31 – 38. Find the slope of the line through the two points.

31.   (–1, 3) and (2, 4)   32.   (–2, 3) and (3, 4)

33.   (–1, –3) and (–2, –4)   34. $(\frac{1}{2}, -3)$ and (–2, 4)

35.   (–1, 8) and (2, –4)   36.   (–4, 3) and (3, 4)

37.   (–2, 3) and $(\frac{1}{2}, -3)$ 38.   (–5, 3) and (3, –6)

✔39. If $P(d) = 0.43d$, evaluate:

    a) $P(100) =$ b) $P(500) =$

    c) $P(w) =$ d) $P(w^2) =$

Complete the following tables of values for each of the following equations. Identify the independent variable and dependent variable in each. Sketch the graph of each set of ordered pairs and identify the slope of any line.

40. $y = 3x - 4$

x	-2	-1	0	1	2	3	4
y							

41. $f(x) = -3x - 2$

x	-3	-2	-1	0	1	2	3
$f(x)$							

42. $y = -\frac{3}{2}x + 2$

x	-2	-1	0	1	2	3	4
y							

43. $y = \frac{1}{4}x - 3$

x	-1	0	1	2	3	4	8
y							

44. $y = \frac{2}{3}x + \frac{5}{4}$

x	-3	0	3	6	9	¼	½
y							

Determine at least two points on each of the following lines and sketch the graph of each line.

45. $y = 2x + 3$ 46. $y = -2x + 1$

47. $y = \frac{1}{2}x + \frac{5}{2}$ 48. $y = -x - 1$

49. $y = x$ 50. $y = -x$

Let $y = 0$ and solve each equation for x to find the x-intercept.

51. $y = 2x - 3$

52. $y + \frac{1}{2} = \frac{2}{3}x - \frac{3}{4}$

53. $y - 11 = -x - 4$

54. $y + 7 = \frac{1}{2}x - 3$

55. $y + 4 = 1 - 2x$

56. $y + 3 = -5x$

Let $x = 0$ and solve each equation for y to find the y-intercept.

57. $y = 2x - 4$

58. $2x + 3y = 4$

59. $y = 3x + 1$

60. $3x - 2y = 6$

61. $y = -\frac{1}{2}x - 5$

62. $\frac{1}{2}x - \frac{2}{3}y = 4$

Express the following equations in the slope-intercept form, $y = mx + b$, then determine the slope and y-intercept.

63. $2x - 3y = 4$

64. $\frac{2}{3}y = x - 4$

65. $-3x + 2y = 9$

66. $y = -2(x + 1)$

67. $x = 2y$

68. $x = y$

✔69. Solve: The height of an average male can be determined using the formula, Height = 67.97 + (1.82)·(length of the arm). How long would the arm be of a man who is 172 cm tall?

✔70. How does the perimeter of a square change if the length of the side is doubled?

Sketch the graph of each line.

71. $4x - 4y = 1$ 72. $-3x + 2y - 7 = 0$

73. $6y = -2x + 4$ 74. $-2x - 4y + 6 = 0$

75. $y = \dfrac{-2}{3}x - \dfrac{7}{3}$ 76. $4x + y = 3$

77.   Study the pattern in each of the following and write an equation for the linear function.

a)

x	1.2	1.3	1.4	1.5
y	13.4	13.1	12.8	12.5

b)

x	2	2.5	3	3.5
y	1.5	2.63	3.76	4.89

c)

x	5	7	10	15
y	35	33	30	25

d)

x	0	-7	-14	7
y	3	6	9	0

✔78. How does the circumference of a circle change if the length of the diameter is doubled?

Write the equation of the line:

79. through the points (9, 4) and (3, 5)

80. through the points (9, 7) and (4, 0)

81. through the points (2, 5) and (0, 9)

82. through the point (8, 6) with a slope of –1

83. through the point (0, –3) with a slope of $\frac{1}{2}$

84. through the point (–3, 0) with a slope of –2

✔85. How does the circumference of a circle change if the length of the radius is doubled?

✔86. If $P(x) = 2x - 3$, then find:

a) $P(0) =$ b) $P(-1) =$
c) $P(1/2) =$ d) $P(4) =$

✔87. If $R(s) = s^2 - 3s + 4$, then find:

a) $R(2) =$

b) $R(-4) =$

c) What is R when s is $\frac{1}{3}$?

d) For what value of s is $R = 4$?

⌘88. Without calculating the slope, determine the sign of the slope of each of the following lines. (Assume a scale of 1 unit for the x-axis and 2 units for the y-axis.)

A)

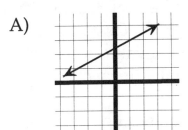

B)

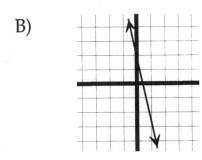

C)

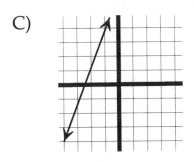

D)

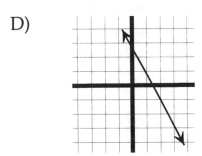

a)   Without actually calculating the slope, determine which line has the most positive slope and which line has the most negative slope.

b)   Verify your answers in part (a) by finding the numerical value of each slope. Label the two points you used to determine the slope with the coordinates of the point.

c)   Write the equation of each line.

89.   Professor Rivera has to make trips to other community colleges for meetings. He is reimbursed for travel at the rate of $0.25 per mile. Write a formula to calculate the amount he is reimbursed as a function of the number of miles he drives. Identify the variables. If he travels over the lunch hour, he is reimbursed $6 for lunch. Adjust your formula to include the cost of lunch. Compare and contrast the graphs of the two functions.

90.   Look up the price of a first-class stamp in 1968, 1971, and 1974. Does the relationship between year and price seem to be linear? If so, use the data to find an equation where price depends on year, and graph the equation. (Hint: Let 1968 be year 0, 1971 be year 3, etc.) Use your equation and/or graph to predict what the price of a stamp was during 1978, 1985, and 1995. Was the predicted value accurate? Is the relationship between year and price actually linear?

91.   Paul earns $6.25 per hour for the first 40 hours worked per week, and earns time and a half for overtime. Write an equation which shows total wages depending on total hours worked in a week with overtime, and use the equation to determine how many hours Paul must work to earn $330.00 before taxes.

92.   Information from the 1995 Federal Income Tax Rate Schedule is given below.

Schedule X – Use if your filing status is Single

If the amount of Form 1040, line 37, is: Over –	But not over –	Enter on Form 1040 line 38	of the amount over –
$0	$23,350	...... 15%	$0
$23,350	$56,550	$3,502.50 + 28%	$23,350
$56,550	$117,950	$12798.50 + 31%	$56,550

a)   If your taxable income is $15,000, what is your tax?

b)   If your taxable income is $40,000, what is your tax? What percentage of your income is the actual tax?

c)   Sketch a graph showing tax as a function of income.

d)   Add a line to your graph that could represent the rate you found in part (b).

93.   List five "real world" examples of linear relationships, other than the ones listed in this unit.

⌘94.   Size of the Slope

A sports company is planning to manufacture a new running shoe which will sell for $124.99 a pair. The shoes cost $47.53 to produce, and the equipment needed to produce these shoes costs $80,000.

a)   What would be your guess about the slope of this line (i.e., what is the additional cost for producing one additional pair of shoes)?

b)   Find an equation which represents this situation and graph it.

c)   Now, suppose the production equipment is depreciating at a constant rate of $1500 per year. Does this describe a linear function? Why or why not?

d)   Find an equation representing this situation and graph it. Does the graph look the way you expected it to look?

e)   Compare the slope of the cost graph with the slope of the depreciation graph. What do you notice?

f)   Make a generalization about how the sign of the slope of a line relates to the graph of the equation of the line.

95. Which of the following represent linear functions. Write the equation of any line.

a)

x	2	4	6	8	10
y	2.5	1.5	0.5	–0.5	–1.5

b)

x	2	4	6	8	10
y	2	14	34	62	98

c)

x	0	5	10	15	25
y	–2	13	28	43	73

d)

x	1/2	3/2	5/2	7/2	9/2
y	–4	–6	–8	–10	–12

96. What is the equation of the line containing all points whose x-coordinate is 3?

97. What is the equation of the line containing all the points on the y-axis?

98. What is the equation of the line containing all the points on the x-axis?

⌘99. Height versus Arm Span
Measure the height and arm span of 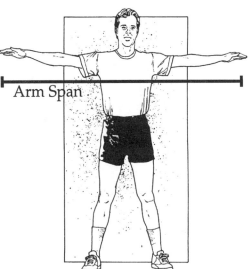 each member of your group. Organize this information in a table. Plot the information on a graph of height versus arm span. Determine the equation of the line which seems to best suit your data and use the graph and/or equation to answer the following questions.

Arm Span

a) About how tall is a person with an arm span of 63 inches?

b) Approximately what is the arm span of a person who is 6 feet 4 inches tall?

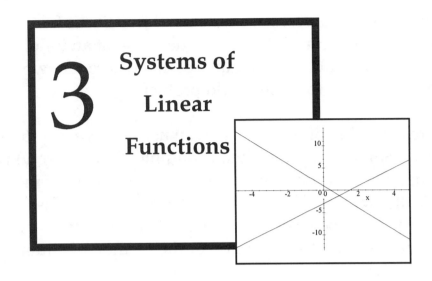

3 Systems of Linear Functions

Upon successful completion of this unit you should be able to:

1. Determine whether two lines are parallel or perpendicular;

2. Approximate solutions of linear systems using graphical and numerical techniques;

3. Solve a system of two equations in two unknowns algebraically; and

4. Solve application problems using systems of equations.

Systems of Linear Functions

In this unit we will investigate systems of two linear functions. We will learn that two lines may intersect at one point, at an infinite number of points, or at no points. The point or points where the two lines intersect may be the solution to a real-world problem.

There are a number of methods we can use to solve a system of two linear equations. Solving a system means finding all points (x, y) which satisfy both equations. Graphically, we can locate the point of intersection. Numerically, we can also use the TABLE feature of the TI-82 graphics calculator to help locate or to approximate the solution. Algebraically, we can use the method of substitution or the method of elimination.

The Coffee Shop

The owner of a gourmet coffee shop has fixed overhead costs of $1500 a month for her shop. The average cost of each cup of coffee is 33 cents. Money she earns selling coffee is her revenue.

1. Write the total monthly cost C to produce x cups of coffee.

2. Find C(40). What does this represent?

3. If she sells coffee for $1.20 per cup, write a function for the revenue R from selling x cups of coffee.

4. Find R(495). What does this number represent?

5. What are reasonable values for the cost function C and the revenue function R?

6. Without actually drawing the graphs of the cost and revenue functions, how do you know the lines will eventually intersect?

7. Next, we will use the TABLE feature of the TI-82 graphics calculator to help locate the solution. Using the TABLE feature will also help you graph the functions later. Enter the cost function into Y₁ using the Ⓨ= key. Then enter the revenue function into Y₂.

8. Press ⓶ⁿᵈ WINDOW to access TblSet (Table Setup) as shown.

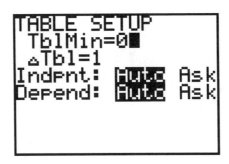

9. TblMin is the minimum value of x you are interested in. ΔTbl is the difference between successive x-values. Try a TblMin of 0 and a ΔTbl of 100 for a first attempt. Now press ⓶ⁿᵈ GRAPH.

10. Press ⏷ until your screen matches the screen below. Notice Y₁ is larger than Y₂ until x is 1800. This means the solution is between 1700 and 1800. Why?

X	Y₁	Y₂
1200	1896	1440
1300	1929	1560
1400	1962	1680
1500	1995	1800
1600	2028	1920
1700	2061	2040
1800	2094	2160

X=1800

11. Next change your TblMin and ΔTbl to locate a smaller interval containing the solution. For example, you might set TblMin to 1700 and your ΔTbl to 10. This means the solution is between _____ and ____. Why?

12. Continue this process until you have found a solution that is correct to the nearest cent.

13. Graph both the cost and revenue functions on your graphing calculator. Be sure the point of intersection appears. For later reference, record the window you used on the TI-82.

[_____ , _____]$_x$ and [_____ , _____]$_y$

14. Sketch the lines.

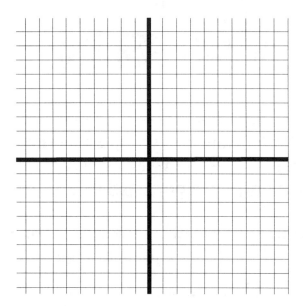

15. Use TRACE to estimate the coordinates of the point of intersection. ZOOM to obtain the solution correct to the nearest cent.

16. What is the significance of the intersection point, and what do the coordinates of the point tell you?

17. Compare the answer to number 14 with the answer to number 12.

18. How many cups of coffee must she sell on an average day to break even? Assume there are 24 business days per month.

19. If she sells 40 cups per day, what will be her profit or loss? What if she sells 75 cups per day?

You Try It

Your Student Government Association (SGA) is planning a dance. The SGA will have to pay a cleaning fee of $75 for the room, and you estimate refreshments will cost $3.75 per person. If the SGA plans to charge $5.50 per ticket, how many tickets will you need to sell to break even?

1.   Write the total cost as a function of the number of tickets sold.

2.   Write the total revenue as a function of the number of tickets sold.

3.   Use the calculator TABLE to estimate the break-even point, the point where cost equals revenue.

4.   Graph both functions on the same coordinate system. Your graph should look similar to the graph below.

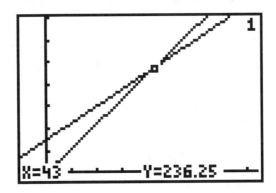

5.   Use the graph to estimate where the cost is equal to the revenue. How many tickets is this?

6.   What are the possible disadvantages of this method?

The major disadvantage of using graphical techniques is that they guarantee only a close estimate, not an exact solution. You should use algebraic techniques when you need an exact solution. You will learn some of these in the remainder of this unit.

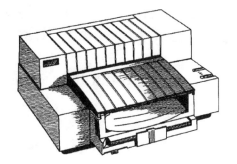

Beyond Graphical Methods

If we develop algebraic techniques for solving systems of linear equations, we can use them to solve application problems.

It takes 4 hours for a printer manufacturer to assemble the parts for its standard printer plus 2 hours to install them in the shell. The top-of-the-line printer requires 5 hours to assemble the parts and 1.5 hours to install them in the shell. The Electronics Department assembles the parts while the Finishing Department installs the parts. Each day the Electronics Department has 200 hours of labor available and the Finishing Department has only 80 hours. How many of each model can they make per day?

If we consider s to be the number of standard models and let t be the number of top-of-the-line models, we obtain the following equations:

$$4s + 5t = 200$$
$$2s + 1.5t = 80$$

To solve the system using the **method of substitution**, we must solve one of the equations for one variable. Although you can use either equation and either variable, it is important to choose with care to avoid as much messy algebra as possible. If we choose to solve the first equation for s, we get $s = -\frac{5}{4}t + 50$ or $s = -1.25t + 50$.

Now substitute this value of s into the other equation:

$$2(-1.25t + 50) + 1.5t = 80$$

Solve this new equation for t:

$$-2.5t + 100 + 1.5t = 80$$
$$-1t + 100 = 80$$
$$-t = -20$$
$$t = 20$$

So we know that 20 top-of-the-line printers can be made in one day. However, we don't know how many standard printers can be made. Now substitute the value of t into the original equation.

$$s = -1.25t + 50$$
$$s = -1.25(20) + 50$$
$$s = 25$$

The manufacturer can make 25 standard and 20 top-of-the-line printers in one day.

You Try It

Solve the following problems using the method of substitution, then verify your solutions using your calculator.

1.
$$3.2x - y = 8.1$$
$$4.5x - 7y = 2.7$$

2. Vince bought 32-cent stamps for letters and 20-cent stamps for postcards. He bought 79 stamps for twenty dollars. How many of each type of stamp did he buy?

An *Interesting* Problem

A second algebraic method is the **method of elimination**. In this method you eliminate one of the variables by combining the terms of the two equations through addition or subtraction.

Sharon has $6000 to invest. She plans to invest part in a certificate of deposit (CD) paying $7\frac{1}{2}\%$ interest and the remaining in risky junk bonds paying 10% interest. If Sharon wants to make $500, how much money should she place in each investment?

Let: c be the number of dollars invested in the CD.
b be the number of dollars invested in junk bonds. So

$$c + b = 6000$$
$$0.075c + 0.10b = 500$$

To clear the decimals from the second equation, multiply each term of the second equation by 1000. Why?

$$c + b = 6000$$
$$75c + 100b = 500000$$

Now, work on the equations so that in each equation one of the variables has exactly the opposite coefficient. To do this easily you might multiply each term in the first equation by −100.

$$-100c - 100b = -600000$$
$$75c + 100b = 500000$$

Next add the left-hand side of each equation together and add the right-hand side of each equation together to eliminate one of the variables.

$$-25c = -100,000$$
$$c = 4,000$$

To achieve her goal, Sharon should put $4,000 in the certificates of deposit. How much should Sharon put into the risky junk bonds?

You Try It

At the basketball game the other night, I looked up and noticed that Jason's foul shot average was 78%. Later after Jason had attempted 2 foul shots, missing one; his average was 76%. How many foul shots has Jason attempted this season?

Parallel and Perpendicular Lines

✎ Beginning with the line below, construct a line with a different
   y-intercept and the same slope. What do you notice about these lines?
   Compare your answer with others' answers and draw a conclusion.

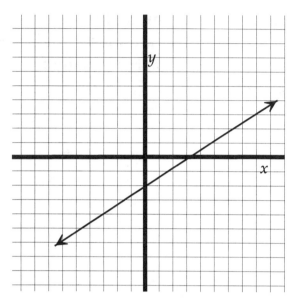

✎ Using the following page of large graph paper, place the corner of a 4"
   by 6" index card or a sheet of paper on the graph paper so the corner of
   the card is on the intersection of two grid lines. Label this point A.

   Keep the corner on A, and move the card so that one side of the card
   lines up with another intersection point of two grid lines.

   Draw a line along each edge of the card. These lines should form a right
   angle. Now calculate the slope of each line. Repeat this experiment
   until you notice a pattern, and make a generalization about
   perpendicular lines.

In the last two experiments you discovered that *parallel lines* always have the same slope and that perpendicular lines have slopes that are related by the following formula: $m_1 = \frac{-1}{m_2}$ or $m_1 \times m_2 = -1$. We often say that slopes of *perpendicular lines* are negative reciprocals of one another.

> *Parallel lines* always have the same slope.
>
> *Perpendicular lines* have slopes related by the formula:
>
> $$m_1 = \frac{-1}{m_2} \text{ or } m_1 \times m_2 = -1$$

You Try It

Determine whether each pair of lines is parallel, perpendicular, or neither.

1. $5x + 4y = 8$

   $y = \frac{-5}{4}x - 3$

2. $8x + 5y = 3$

   $8x - 5y = -3$

3. $2x + 3y = 6$

   $3x - 2y = -3$

4. $275x - 214y = 947$

   $y = \frac{275}{214}x - 345$

5. Write the equation of a line parallel to $y = -0.4x - 2.3$ passing through the point $(2, 3)$.

6. Write the equation of a line perpendicular to $y = -2x + 5$ passing through the point $(-1, -2)$.

Inconsistent Systems of Equations: Parallel Lines

Two catering concerns are in competition for business. At Jill's Catering, Jill charges $7.95 for each Teriyaki Chicken meal. At Cory's Chicken on Wheels, Cory charges $7.95 for each Teriyaki Chicken meal plus a delivery charge of $20. When will Jill and Cory have brought in the same amount of money for selling exactly the same number of meals?

After writing an equation for each situation we find that Jill's revenue will be $R_j(x) = 7.95x$ where x is the number of meals sold. Cory's revenue will be $R_c(x) = 7.95x + 20$. Let $R_j = R_c = y$. If we solve this system algebraically using the method of elimination, we find:

$$
\begin{aligned}
y &= 7.95x \\
-y &= -7.95x - 20 \\
0 &= -20
\end{aligned}
$$

The last equation is, of course, false. At this point you should graph each of the equations. The lines appear to be parallel. How can you convince other class members the lines are parallel?

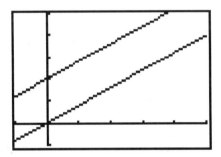

If a system of two equations produces a graph of parallel lines or a false statement when you solve it algebraically, it has no solution. This is called an **inconsistent system**. If the lines intersect at only one point, it has a unique solution and is called a consistent system. Remember, parallel lines will never intersect.

You Try It

1.   If possible, solve the system:

$$4x - 8y = 10$$
$$-3x + 6y = -2$$

What is the slope of each line? What are the y-intercepts?

2.   If possible, solve the system.

$$0.52x + 1.99y = 5.86$$
$$0.29x - 0.92y = 8.8$$

What is the slope of each line? What are the y-intercepts?

Dependent System of Equations: Same Line

Kate started a rigorous exercise program. On the first day she walked at a steady rate for half an hour and then rode a bicycle for another half an hour, also at a constant rate. She covered 6 miles. The next day, Kate tried to maintain the same walking speed and biking speed as the day before, but she walked for one-third of an hour and rode for twenty minutes covering 4 miles total. What were her speeds walking and biking?

If we let the rate at which she walked be x and the rate at which she rode be y, we can write two equations to represent the two days. We converted all times to minutes.

$$30x + 30y = 6$$
$$20x + 20y = 4$$

If we multiply each term in the first equation by –2 and each term in the second equation by 3 so that the coefficients of each x are opposites, we obtain

$$-60x - 60y = -12$$
$$60x + 60y = 12$$

Adding terms, we find $0 = 0 + 0$ which, of course, is always true.

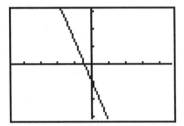

When we graph the two equations, there appears to be just one line. This is called a **dependent system of equations**.

Every ordered pair on the first line is also on the second line. Each point on the line is a solution to both of the equations. There are an infinite number of combinations of walking speeds and biking speeds that would provide the same number of miles. For example, she could walk at one mph and bicycle at 11 mph, or she could walk at 2 mph and bicycle at 10 mph. In each case she would still cover 6 miles in the hour.

You Try It

1. If possible, solve the system:

$$3x - 2y = 1.5$$
$$1.5x = 0.75 + y$$

What is the slope of each line? What are the intercepts?

2. If possible, solve the system:

$$9x + 8y = 64$$
$$y = -1.125x - 27$$

What is the slope of each line? What are the intercepts??

Summary

I.   You have seen examples of four ways to solve systems of linear equations:
- graphically;
- numerically using the TABLE feature of the calculator; and
- algebraically using;
  - ◦ the method of substitution and
  - ◦ the method of elimination;

Many different points are the correct solution to a single line. Two lines may intersect at one point, at an infinite number of points, or at no points. The point or points of intersection is the solution to the system of equations. We may determine this point algebraically by elimination or substitution. We can also estimate the point using the TABLE and TRACE features of the TI-82.

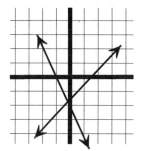 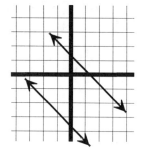 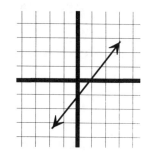

Consistent
One Solution
The two graphs intersect at one point.

Inconsistent
No Solution
The two graphs are parallel and do not intersect. There is no solution to the system.

Dependent
Infinitely Many Solutions
The graphs of each equation produce the same line. Each point is a solution to the system.

A system of equations has no solution if it produces a false statement (such as $3 = 7$ or $0 = 4$) when solved algebraically. If the system produces a true statement (like $0 = 0$ or $2 = 2$) when you solve it algebraically, it has an infinite number of solutions. Remember the solution occurs when the lines intersect.

II. Parallel and Perpendicular Lines

- Parallel lines always have the same slope.

- Perpendicular lines have slopes which are related by the following formula: $m_1 = \frac{-1}{m_2}$.

ProCats

Memorandum #3   **Lexington, KY 40506**

TO: Team Members

FROM: Your Supervisor

Management is considering opening a new souvenir shop that will sell T-shirts. The T-shirts cost $9.50 and will be sold for $19.95. The cost of operating the shop will be approximately $800 per month. Management needs the team's assistance in deciding whether or not to open the shop. Please complete the situation analysis and report as outlined below.

The Analysis:

1. Construct a monthly cost function.

2. Construct a monthly revenue function.

3. Graph both functions.

4. Determine the break-even point algebraically and use your graph for confirmation.

5. Construct a monthly profit function.

6. Graph the monthly profit function and construct a table of values which might be useful to management.

The Report:

Write a report to submit to your supervisor which includes the requested information and conclusions. The report should explain your work, your analysis, any assumptions you made, and comments about the reliability of your results as well as your conclusions.

Problems for Practice

A. Use the TABLE feature of the calculator to approximate the solutions to the nearest tenth, then graph and trace to verify.

1. $9x + 4y = 35$
   $9x + 7y = 40$

2. $9x - 6y = -8.8$
   $3x + 1y = 8.8$

3. $x - 4y = 6.6$
   $7x + 8y = -2.9$

4. $-5x - y = 4.7$
   $x - 0.9y = 9.5$

B. Trace to estimate the solutions of the following linear systems. Then solve by substitution.

5. $y = 3x$
   $2x + y = 10$

6. $x + 3y = 25$
   $x = 2y + 5$

7. $2x + y = -10$
   $6x - 3y = 6$

8. $x + 5y = -9$
   $4x - 3y = -13$

9. $x - y = 4$
   $3x - 3y = 10$

10. A sporting goods company can produce a basketball for $12. The daily fixed costs are $720, and the company plans to sell each basketball for $20.

   a) Graph the cost and revenue equations on the same coordinate system and find the break-even point.

   b) Record the WINDOW: [_____ , _____]$_x$ and [_____ , _____]$_y$

   c) What is the profit or loss if 200 basketballs are sold in a day?

   d) What is the profit or loss for 300 basketballs?

   e) How many basketballs must they sell to produce a daily profit of $200?

C. Trace or use the calculator TABLE to estimate the solutions of the following linear systems, then solve using the elimination method.

11. $-3x + y = 15$
     $2x - y = 10$

12. $3x - 8y = -16$
     $7x + 8y = -14$

13. $y = 3x - 2$
     $3x - 2y = 28$

14. $2x + y = 6$
     $3x - 4y = 12$

15. $x - y = 4$
     $3x - 3y = 12$

16. Jeremy and Andrea want to put a rectangular flower bed in front of their house. The house is 30 feet wide and has a driveway beside it. They plan to place the flower bed with one side against the house and the other against the driveway so there will be less grass to trim. They bought 10.5 feet of edging to put along the other two sides of the flower bed. They think the flower bed will look better if it is twice as long as it is wide. What dimensions should they use?

D. Use any method to solve the following systems of equations. In each case, explain why you chose the method you used.

17. $3x - y = -5$
     $2x + y = 4$

18. $2s + 3t = -3$
     $3t - 2s = 2$

19. $3a - b = -5$
     $2a + 3b = 4$

20. $3s + t - 3 = 0$
     $2s - 3t - 2 = 0$

21. $\frac{3}{2}x + 4y = -3$
     $2x - \frac{2}{3}y = 3$

22. On Rojelita's last hiking trip, she covered a total distance of 16 miles in 6 hours. It seemed to her she was climbing hills most of the time. If Rojelita estimates she walks at a rate of 2 mph climbing hills and 3 mph otherwise, how much time did she spend climbing hills?

23. At Cumberland Falls the group walked 0.5 of a mile. They noticed a sign that said they had gone up 400 feet. If there are 5,280 feet in a mile, what would be the slope of the hill?

24. The owner of McKirnan Brothers wants to introduce a special candy mixture of chocolate-covered pretzels and chocolate-covered peanuts. The plan is to mix 12 pounds of the mixture at a time to be sold for $3.00 a pound. Chocolate-covered pretzels sell for $2.75 a pound while chocolate-covered peanuts sells for $3.50. How many pounds of each should be in the mixture?

25. It has been determined that a temperature in Celsius varies directly as the difference between the Fahrenheit reading and $32°$ F. If a Celsius thermometer reads $37°$ when the Fahrenheit one reads $98.6°$, what will be the temperature in degrees Celsius when the Fahrenheit thermometer reads $5°$.

26. A company manufactures both CD players and VCRs. The cost of material for a CD player is $20, and the cost of materials for a VCR is $80. The cost of labor to make a CD player is $30; to make a VCR is $50. During a week when the company budgeted $4200 for materials and $2800 for labor, how many CD players did the company plan to manufacture?

E. Calculate the slopes of the following pairs of lines and state whether the pair of lines is parallel, perpendicular, or neither.

27. $2x + y = 4$
$-2x - 3y = 9$

28. $x + 2y = 5$
$2x - y = 9$

29. $2x + 3y = 4$
$3x - 2y = 11$

30. $y = 3x - 9$
$-2x + \frac{2}{3}y = 4$

F.   31.   Write the equation of a line parallel to $y = 3x - 4$ passing through the point $(-3, 2)$.

32.   Write the equation of a line parallel to $-3x + 2y = 14$ passing through the point $(17.2, -10)$.

33.   Write the equation of the line perpendicular to $2x - 4y = 8$ passing through the point $(9, 7)$.

34.   Write the equation of the line perpendicular to $7x + 8y = 6$ passing through the point $(-1, 1)$.

35.   On graph paper, draw a line segment whose endpoints have the coordinates $(-2, 1)$ and $(6, 7)$. Use your knowledge of the slopes of perpendicular and parallel lines to draw a rectangle. State the slopes of the 4 sides. For each of the four sides, write the equation of the line containing the side.

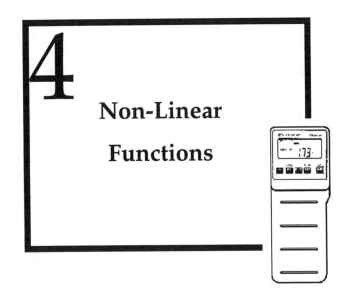

Upon successful completion of this unit you should be able to:

1. Determine if a graph represents a function;

2. Identity the domain and range of a function;

3. Connect a graph with a verbal representation;

4. Write an equation for a situation involving inverse variation;

5. Compare power and exponential functions; and

6. Describe at least three non-linear functions from everyday life.

Many real-life situations can be modeled using non-linear as well as linear functions. Some of these are represented by exponential functions, power functions, and inverse variation.

This unit will present more properties of functions and introduce some new functions. You will learn to test whether a graph represents a function and be able to find its domain and range. In addition, you will create graphs from real-world situations.

Light Intensity and Distance

Do you remember when your mother used to tell you not to read in the dark, but to sit under the light? How does the brightness of the light change as you move closer or farther from the light source?

We can use the Calculator Based Laboratory (CBL) with the TI-82 to examine the relationship between light intensity and distance. The data collected by the CBL can be stored in lists and then plotted using the TI-82.

Before performing the experiment, predict the type of graph you expect from the data collected. Will it be a line or do you expect something else?

Equipment Needed:

- CBL Unit
- TI-82 graphics calculator with a link cable
- TI light probe
- Standard light bulb (15 or 25 watt)
- Tape
- Wooden Block
- Meter Stick

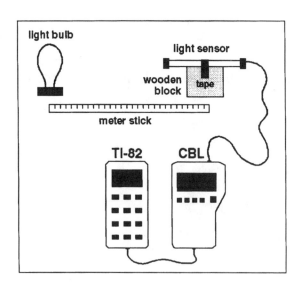

Instructions:

1. Use tape to secure the light probe to a wooden block. The light probe should face the light bulb.

2. Start the LIGHT program provided by your instructor on the TI-82. Enter the number of data points you will collect.

3. Position the light probe so it is 10 centimeters from the center of the bulb. After you collect the light intensity for this position, move the light probe 10 centimeters farther from the center of the bulb. Repeat this process until you have collected the number of data points you indicated.

4. After the data has been collected, the graphing calculator will display a plot of light intensity (mW/cm^2) versus separation (centimeters).

Analysis

Here is the graph screen from one run of the experiment. Does this look like the one from your class?

Notice your graph does not represent a line. Later in the chapter you will learn that light intensity versus separation is an example of **inverse variation**. The general function representing light intensity versus separation is $W(s) = \dfrac{c}{s^2}$ where c is a real number.

Exploring the Domain and Range of Functions

Mathematicians have always had a curiosity about and studied the relationships between sets of numbers. Some of the relationships are highly complex while others are interesting yet simple enough for anyone to study.

We are often interested in finding a relationship between two sets of numbers. For example, you might keep track of the gallons of gasoline you purchase when you drive so many miles.

No. of Miles Driven	No. of Gallons of Gas
400	13
336	11.2
299	11.5
200	6.75
157	5.75

This situation pairs the members of two sets according to the efficiency of the engine. Every member from one set called the **domain** is paired with at least one member in a second set called the **range**.

It is perhaps easiest to examine a set of ordered pairs to explore further the concepts of domain and range.

The numbers in the table can be written using the following set of ordered pairs, {(400, 13), (336, 11.2), (299, 11.5), (200, 6.75), (157, 5.75)}.

Plot the points on the WINDOW $[-50, 450]_x$ and $[-1, 14]_y$.

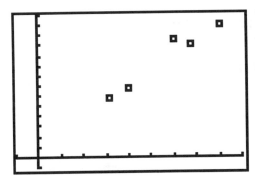

The domain is the set of first coordinates of the ordered pairs, namely {400, 336, 299, 200, 157}.

The range is the set of all second coordinates of the ordered pairs, or {13, 11.2, 11.5, 6.75, 5.75}.

Recall from Unit 2:

> A **function** is a relationship between two variables (one independent variable and one dependent variable) where independent variable value is paired with exactly one dependent variable value.

A function pairs an independent x (a number from the domain) with a dependent y (a number in the range). The definition can be also stated as:

> A **function** can also be defined as a *rule* that assigns to each value in the **domain** exactly one value in the **range**.

A function can be represented by an equation, a graph, a set of ordered pairs, a table, or in words.

You Try It

Determine which of the following are functions and explain why.

1.　　For each course a student takes there corresponds one final grade.

2.　　For each student there may correspond several courses.

3.

a	b
-1	-3
-2	-2
0	-1
1	-3
2	1
1	-3
-1	-3

4.

c	d
-1	-2
-2	-3
0	-4
1	0
2	3
1	2
-1	4

5.

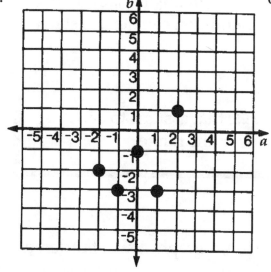

6.

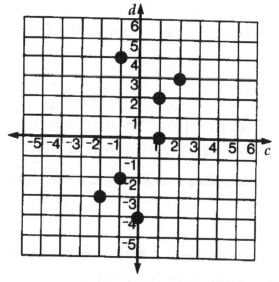

7.　　Find the domain and range for any function in numbers 3 – 6. You should apply the definition in the same manner whether you are looking at functions verbally, graphically, or numerically.

Determining the Domain Graphically

It is often easier to determine the domain and range of a function from the graph than from any other representation of the function. To determine the domain, you might consider the following scenario: Imagine it is a bright, sunny day and the sun is shining directly overhead. A point on the graph blocks the sunlight and causes a shadow to fall on the x-axis. To find the domain, you *must* project every point on the given graph onto the x-axis.

Find the domain of the function sketched on the WINDOW, $[-5,5]_x$ and $[-3,5]_y$. The endpoints are $(-2, 2.1)$ and $(4, 1.9)$.

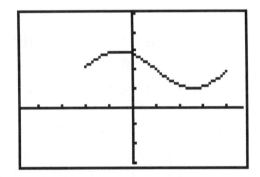

To find the domain, project the graph vertically onto the x-axis as shown.

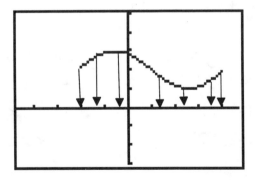

So the domain of the function given by the graph is the set of values of x between –2 and 4 inclusively. In interval notation, we write $[-2, 4]$ and using inequalities, $-2 \le x \le 4$.

More often, the graph has values that continue past the edge of the TI-82 screen. In this text, unless otherwise stated, you can add arrows to the end of any graph that seems to end at the edge of the screen. For example, the TI-82 graph on the left might be sketched on graph paper as the graph on the right.

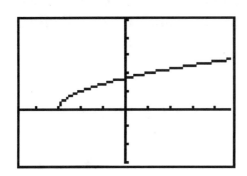 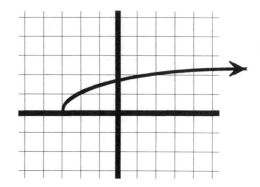

For this function, the possible values for x begin with -3 and then continue on toward infinity, ∞. In interval notation, we write the domain as $[-3, \infty)$. Using inequalities, the domain is $-3 \leq x < \infty$ or $-3 \leq x$.

Domains and ranges are always expressed with the smaller number first. When writing an interval which includes an endpoint, use a bracket. If the endpoint is not included, use a parenthesis. When the interval notation includes infinity (∞), a parenthesis is always the correct symbol.

The chart below provides a comparison of interval notation with the inequality representing the same values.

Interval Notation	Inequality
$[-2, 4.7]$	$-2 \leq x \leq 4.7$
$(-2, 4.7)$	$-2 < x < 4.7$
$(-2, 4.7]$	$-2 < x \leq 4.7$
$[-2, \infty)$	$-2 \leq x < \infty$

You Try It

For each of the domains expressed in interval notation, write the domain using inequalities.

1. $[-3, 8)$

2. $(2, 800]$

3. $(-\infty, \infty)$

For each of the domains expressed using inequalities, express the domain in interval notation.

4. $-400 < x < \infty$

5. $-35.3 \leq x \leq 75$

Range

To find the range, project the graph horizontally onto the y-axis. If we could trace the graph it would be easy to see that the values of the range are all the y values between 1 and 3. However, since we are approximating the range using the graph, any answer reasonably close to those values is a good answer. We can write the range as $[1, 3]$.

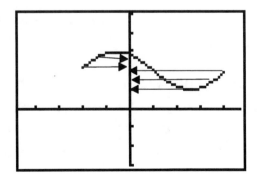

You Try It

State whether the graph represents a function or not. Write the domain and range of the functions in interval notation. Each graph has the following window: $[-7, 6]_x$ and $[-7, 7]_y$.

1.

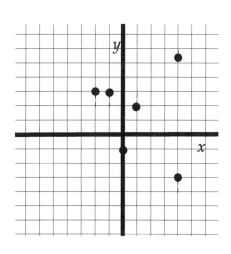

2.

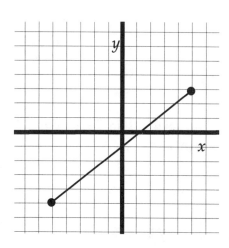

3.

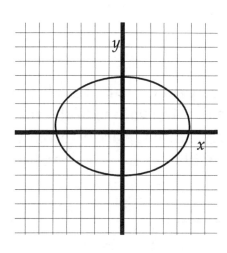

4.

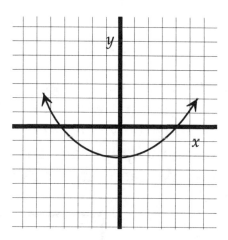

Notice for each graph above that was not a function, there was at least one pair of points you could connect with a vertical line. To determine whether a graph is a function, check to see that no vertical line will intersect the graph at more than one point. This is called the **Vertical Line Test**.

Each vertical line in the coordinate system intersects the graph of a function in no more than one point because a function is a rule that assigns to each value in the domain exactly one value in the range.

Which of the following graphs represent functions?

A

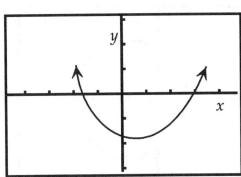

B

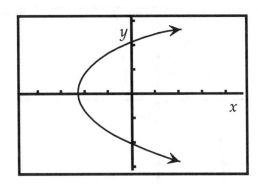

Draw several vertical lines through the graph.

A

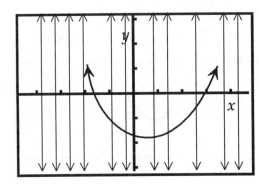

B

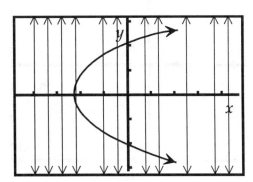

On graph A each vertical line intersects the graph in no more than one point. So graph A is the graph of a function.

Some vertical lines intersect graph B at more than one point. So graph B is **not** the graph of a function.

Your graphing calculator is sometimes called a function grapher because the only thing you can enter in ⟨Y=⟩ is a function.

You Try It

Determine which of the following graphing calculator graphs represent functions graphed on $[-4.7, 4.7]_x$ and $[-3.1, 3.1]_y$. Determine the domain and range of each function.

(1)

(2)

(3)

(4)

(5)

(6)

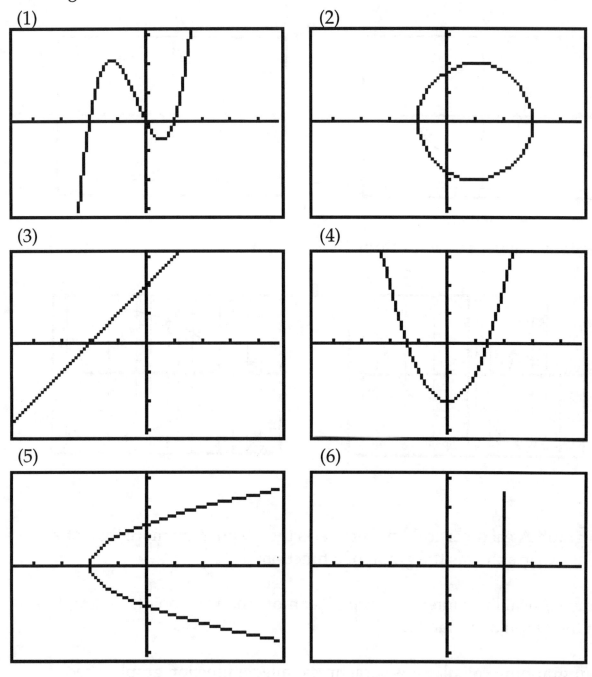

Identifying Qualitative Graphs

To observe the overall features of a function, a graph is invaluable.

Which graph best matches the situation — a train pulls into a station and lets off its passengers? Why?

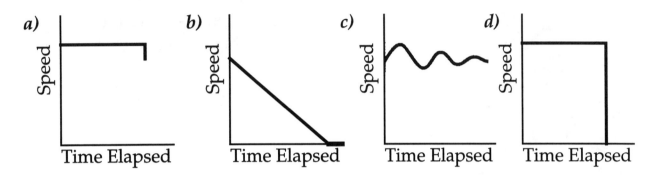

Which of the graphs above represent functions?

You Try It

Which graph best matches the situation — Jane climbs the hill at a steady pace then starts to run down the other side? Justify your answer.

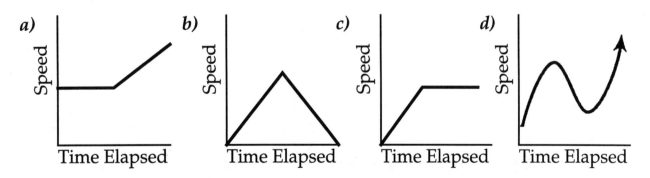

Which of the graphs above represent functions?

Adapted with permission from "Activities: Relating to Graphs in Introductory Algebra," by Frances Van Dyke (*Mathematics Teacher*, Vol. 87, No. 6), copyright September 1994 by the National Council of Teachers of Mathematics.

Functions in Words to Graphs

Many problems become clearer when you sketch a graph.

Sketch a reasonable graph of each of the situations below. Label each axis and specify a reasonable domain and range. State any assumptions you are making.

1.   The distance from the ground to the bottom of the big toe on your left foot as you bicycle.

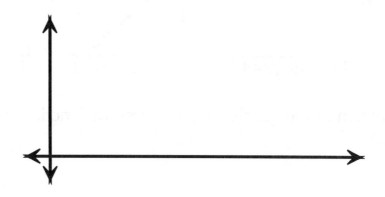

2.   The amount of time you can safely spend in the sun without burning as related to the number of the sunscreen lotion you use.

Functions as a Rule

In Kentucky on a warm summer evening, a familiar sound may be the chirp of a cricket outside your bedroom window. The rate at which the cricket chirps *depends* on the temperature. The warmer the weather, the more the cricket chirps. The table below shows how temperature and rate of chirping are related. Complete the table assuming the number of chirps is a linear function of the temperature. Write the rule and determine the domain and range.

Temperature in degrees Fahrenheit	50	60	70	80	90	100
No. of chirps in 1 minute	10	20	30	40		

To each temperature in the table, there corresponds a rate. We say the rate at which crickets chirp is a function of the temperature.

You Try It

Use the definition of a function as a rule to complete the following:

1.   The length of line needed for the anchor of a boat is a function of the water depth.

Depth of water in meters	3	6	10	13	16
Length of line in meters	21	42	70	91	???

What is the missing number from the table?

How can the length be found from any given water depth?

2. Plot the points and complete the pattern. Is it a function? Determine the domain and range.

a)

x	f(x)
0	5
2	11
-3	-4
3	14
4	
-4	
5	
2	

b)

a	b
3	9
4	16
5	25
6	
7	
-1	
-2	4
-3	

c)

x	y
2	24
3	16
4	
6	8
8	6
-12	-4
-8	
-6	

Which are linear? Which are non-linear?

Variation

Recall that direct variation is linear. If you look back at number 1, the deeper the water, the more line is needed for the anchor.

✎ Circle the best answer for each of the following:

1. The more you study the (better/worse) grade you get.

2. The higher the cost of a car, the (more/less) the sales tax will be.

The answers "better" and "more" illustrate direct variation.

✎ Circle the best answer for the following:

1. The colder the temperature is outside, the (higher/lower) your heating bill will be.

2. The faster you drive the (more/less) time it takes to reach your destination.

The choices "higher" and "less" illustrate inverse variation.

✎ Discuss what you anticipate the differences are between direct and inverse variation. Can you think of another example of direct variation and another example of inverse variation?

Inverse Variation

Whenever a real-life situation produces pairs of numbers whose product is always the same or constant, we say there is **inverse variation**.

Billy noticed when he set the cruise control on his car at 40 mph, the car got 12 miles per gallon (mpg); at 50 mph, the car got 9.6 mpg; at 60 mph, the car got 8 mpg. Notice that as the speed increases, the mpg decreases or *the mpg varies inversely with mph*. Let *s* represent speed and *G* represent mpg.

Billy's Table:

s	40	50	60	70
G	12	9.6	8	??

Note as the first number gets larger, the second number gets smaller. Multiplying each pair of numbers together gives 480.

$$[40 \cdot 12] = [50 \cdot 9.6] = [60 \cdot 8] = 48$$

We can set up an equation that expresses this fact:

miles per gallon × miles per hour = constant

or $\qquad\qquad\qquad\qquad\qquad G \times s = $ 480 (a constant)

> If y varies inversely as x, then $y = \dfrac{k}{x}$, for some constant k.

1. Use the previous table of values for this mileage problem and graph the relationship on paper. (Label axes with appropriate variables.)

2. Did you get a line? Graph the function on the TI-82.

3. Use the TABLE feature of your TI-82 to evaluate the function at different values. At TblSet, choose Ask for Indpnt: as shown. Once you press 2nd GRAPH, enter the *x*-value.

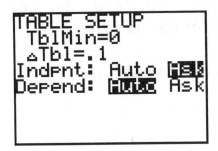
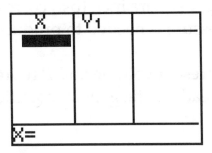

4. What would be the mpg corresponding to 80 mph? What would be the mph corresponding to 20 mpg?

5. Use the TABLE feature to calculate y-values as *x* gets closer and closer to zero. What happens?

6. What are the intercepts?

7. What would happen to the mpg if the speed is doubled? Halved?

8. What is the domain of the function? What is the range? What is the domain of the situation?

As we learned in direct variation, each situation has a different k, but the k is constant as long as the situation does not change. Part of most direct and inverse variation problems is to use the given information to find the constant, k. Once we calculate the k, the value of k is substituted into the general equation, $y = \frac{k}{x}$. Then the equation with only x and y is used to answer specific questions.

Oxygen Tank

At a constant temperature the volume of gas in an oxygen tank varies inversely with the pressure of the oxygen in the tank. If a pressure of 48 pounds per square inch (psi) corresponds to a volume of 50 cubic feet, what pressure is needed to produce a volume of 100 cubic feet?

Let V represent volume and P represent pressure. *Volume varies inversely as pressure* leads to the general equation,

$$V = \frac{k}{P}.$$

Using the given information, $P = 48$ psi and $V = 50$ ft³, we have

$$50 = \frac{k}{48}$$
$$k = 50 \cdot 48$$
$$k = 2,400$$

Now we write the specific equation that represents the relationship between pressure and volume in this situation.

$$V = \frac{2400}{P}$$

Next, answer the question by substituting $V = 100$ into the equation above.

$$100 = \frac{2400}{P}$$
$$100P = 2400$$
$$P = 24 \text{ psi}$$

In this oxygen tank, a volume of 100 cubic feet is produced by a pressure of 24 pounds per square inch.

✎ What pressure is needed to produce a volume of 237 cubic feet?

✎ What volume is produced by a pressure of 78 psi?

You Try It

Peach, Inc., producer of notebook computers, determined that the number of computers it can sell (S) varies inversely with the price (P) of the computer. Two thousand computers can be sold if the price is $2500 per computer.

1. What is the general equation?

2. What is value of the constant k?

3. What is the specific equation?

4. How many computers can they sell if the price is reduced to $2000 per computer?

Light Intensity Revisited

Earlier in the unit you collected data for light intensity versus separation using the CBL. Experiment with different values for c to determine a function that appears to pass through the points if the function $y = \frac{c}{x^2}$ represents the general equation.

Power and Exponential Functions

Recall from an earlier mathematics class that $y = \frac{1}{x}$ can also be written using a negative exponent, $y = x^{-1}$. Thus, the inverse variation problems we have been studying can also be written as $y = kx^{-1}$. The exponent is often referred to as a power. We refer to functions where the variable is raised to a constant power as power functions. Examples of power functions that we will be studying include $y = x^2$ and $y = x^3$.

> A power function has the form $y = f(x) = kx^p$ where k and p are constants.

✎ To graph $y = f(x) = x^2$, complete the table, plot the points, and connect the points with a smooth curve.

x	$f(x)$
−3	9
−2	4
−1	1
0	0
1	1
2	
3	

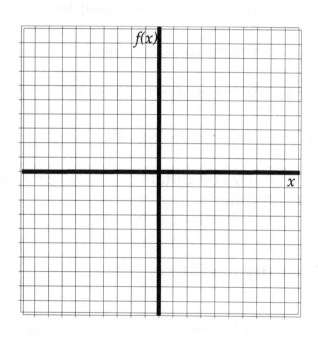

You saw the graph of $f(x) = x^2$ in Unit 1, but now suppose we switch the x and p in the general power function. The new function is an exponential function because the variable is the exponent.

An exponential function has the form $y = g(x) = kp^x$, $p > 0$, $p \neq 1$ where k and p are constants.

✎ Graph the exponential function $g(x) = 2^x$. Complete the table, plot the points, and connect with a smooth curve.

x	$g(x)$
–3	0.13
–2	0.25
–1	0.5
0	1
1	2
2	
3	

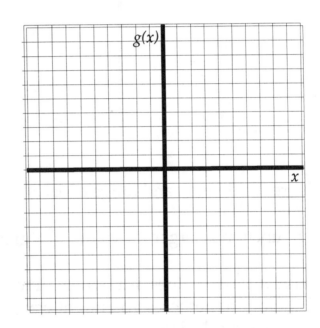

To see more of each graph, graph both $y = x^2$ and $y = 2^x$ on your graphing calculator.

✎ How are these graphs similar? In what ways are these graphs different? What are the domains of the functions?

✎ Which graph has no x-intercept? Are the y-intercepts the same points?

✎ Compare the range of $y = x^2$ with the range of $y = 2^x$.

Graph $f(x) = x^3$ and $H(x) = 3^x$ on this coordinate system.

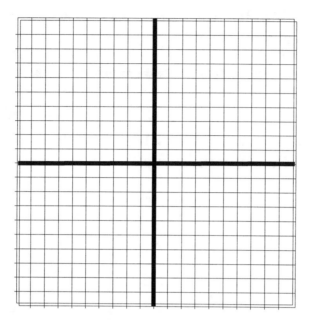

1. What are the domain and range of each function?

2. Which function is larger for values of x greater than 5? Less than –2?

3. For large positive values of x, which function has larger y values? To confirm your hypotheses you might want to use the TABLE feature on your graphing calculator.

Exponential Growth: Analysis of Sales of Videocassette Recorders

During the late 70's and 80's, the sale of VCR's nationwide skyrocketed. Every family wanted to own a VCR to play videos rented at the local video store. The following investigation will allow you to explore exponential growth both numerically and graphically.

The following table shows the number of households with VCR's in millions of units. Complete the table.

Year	Years after 1980	Millions of House-holds with VCR's
1,980	0	1
1,984		9
1,985	5	18
1,986	6	31
1,987		43
1,989		51

Source: *Statistical Abstract of the United States*, 1993.

✎ Plot the points corresponding to Millions of Households with VCR's versus Years after 1980 on your graphing calculator using an appropriate scale.

A function that approximates these values is the exponential function, $v(x) = 0.60(1.9058)^x$. Use this exponential function to answer the following questions.

✎ In 1983, to the nearest tenth, approximately what was the number of households with VCR's?

✎ To the nearest tenth, what does the function predict will be the number of households with VCR's in 1996? Does your answer seem reasonable? Why or why not?

Summary

I.   Functions

- ◆ A function is a rule that assigns to each value in the domain exactly one value in the range. Use the Vertical Line Test to graphically determine whether the graph is a function.

- ◆ The domain of a function is the set of all possible values of the independent variable. The range is the set of all possible values of the dependent variable.

The following chart illustrates different methods of specifying a function.

Method	Example	Domain	Range
A set of ordered pairs	$\{(0,-2), (3,7), (-3,7)\}$	$\{0, 3, -3\}$	$\{-2, 7\}$
A table	 X Y1 -4 14 -3 7 -2 2 -1 -1 0 -2 1 -1 2 2 Y1⊟X^2-2	 X -4 -3 -2 -1 0 1 2	 Y1 14 7 2 -1 -2 -1 2
A graph		$-\infty < x < \infty$, (∞, ∞)	$-2 \leq x < \infty$, $[-2, \infty)$
An equation	$y = x^2 - 2$	$-\infty < x < \infty$, (∞, ∞)	$-2 \leq x < \infty$, $[-2, \infty)$

II.   Variation

- If y varies directly as x, then $y = kx$, for some constant k.
- If y varies inversely as x, then $y = \dfrac{k}{x}$, for some constant k.

III.   Power Functions

A power function has the form $y = f(x) = kx^p$ where k and p are constants.

IV.   Exponential Functions

An exponential function has the form , $y = g(x) = kp^x$ with $p > 0$ and $p \neq 1$ where k and p are constants.

This unit has introduced more properties of functions and some non-linear functions. The next unit will explore an important non-linear function in-depth.

Supplies Needed: Regular m & m's
                        Small paper cups (bathroom size)

Procedure:

Trial #	# of m&m's left
0	
1	
2	
3	
4	
5	
6	
7	

a. Count & record the number of m & m's you received. This is trial number 0.
b. Put the m & m's back in the cup.
c. Pour out the m & m's in a single layer.
d. Remove any m & m's that do not land "m" up.
e. Record the trial number and number of m & m's remaining.
f. Put the remaining m & m's back in the cup.
g. Repeat steps c through f until no m & m's remain.
h. Plot the points above on your graphing calculator using a line graph plot. Do *not* plot any points where the number of m & m's left is zero.
i. Examine the table for patterns.
j. Compare your plot to a line. Is there a constant difference?
k. Compare your plot to an inverse variation graph. Is the product of the Trial # and the # of m & m's left a constant?
l. Approximately what fractional part of your m & m's from Trial #2 remain in Trial #3? About what fraction of your m & m's from Trial #3 remain in Trial #4? Continue this process and determine if a pattern exists.
m. Finally, compare your graph to an exponential graph of the form $M(t) = c\left(\frac{1}{2}\right)^t$ where c is a constant. Experiment to determine a constant, c, so the function appears to pass through most of the points?

Problems for Practice

A.   State whether the graph is a function or not. Find the domain and range of each function. Unless otherwise specified, the window for each is $[-8, 8]_x$ by $[-8, 8]_y$.

1.

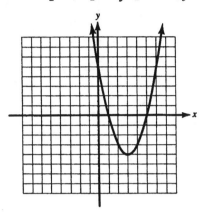

2.

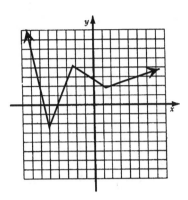

3.

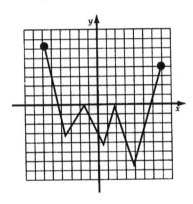

4.

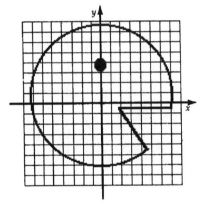

5.

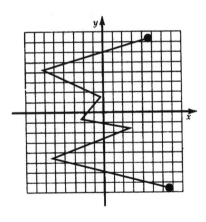

6.

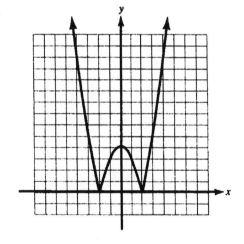

Use $[-8, 8]_x$ by $[-2, 14]_y$

B. Use the Vertical Line Test to determine which of the following are graphs of functions.

7.

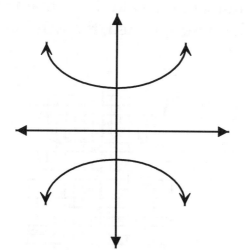

8

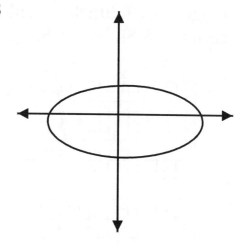

9.

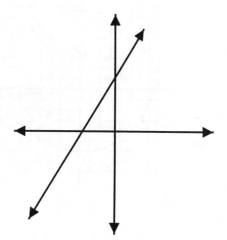

10

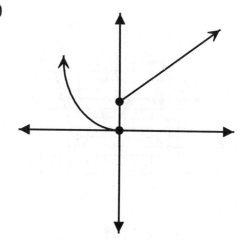

11.

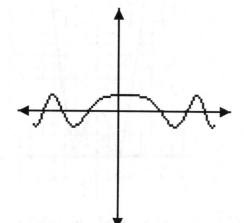

12.

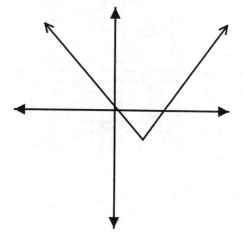

C. Indicate which graph best matches the statement. Which of the graphs below do <u>not</u> represent functions?

13. A child swings on a swing.

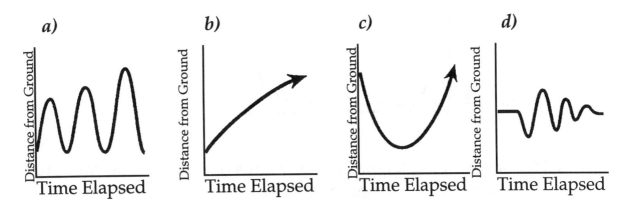

14. A child climbs up a slide then slides down.

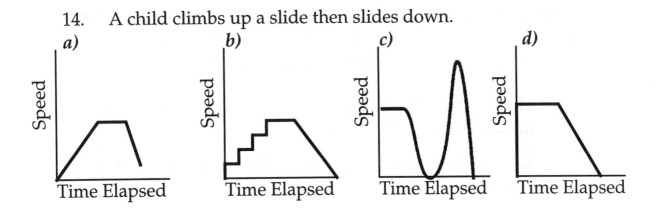

15. You dive into the water from the diving board.

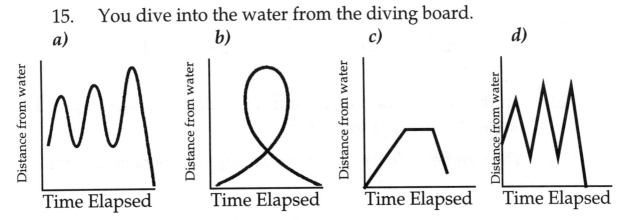

Adapted with permission from "Activities: Relating to Graphs in Introductory Algebra," by Frances Van Dyke (*Mathematics Teacher,* Vol. 87, No. 6), copyright September 1994 by the National Council of Teachers of Mathematics.

16. Explain to your friend why you could never have a graph like the one sketched below.

✔17. Graph circumference versus radius of a circle. Use your graph to predict the circumference of a circle with a radius of 7.3 meters.

D. Sketch a reasonable graph of each of these situations. Label each axis. Specify a reasonable domain and range, and state any assumptions your group is making. Finally, is it a function?

18. The number of hamburgers sold by McDonald's as compared to the age of the corporation

19. The distance your group is from the band and how loud it sounds

20. The number of hours of daylight from January 1 to December 31

21. The winning speed in furlongs per hour at the Kentucky Derby from 1960 to 1980

22. The body temperature of a murder victim dependent upon the length of time after the murder.

23. The number of kernels of popcorn popping as related to time

E. Write a variation equation for the following using k as the variation constant.

24. I varies inversely as R.

25. a varies inversely as the cube root of b.

✔26. C varies directly as r^2.

F. For the following, identify whether the variation is direct or inverse, and find the domain and range.

27. $a = \dfrac{2.5}{d}$

28. $c = 4f$

29. $b = 270w$

30. $t = \dfrac{3600}{h}$

G. For each of the following:

a) Sketch the graph.
b) Classify the function as representing direct or inverse variation.
c) Find the formula for the relationship.
d) Use the formula to find two additional values for each table.

31.

x	2.5	7	16
y	8.25	23.1	52.8

32.

x	2.5	7	16
y	1.32	0.47143	0.20625

33.

x	4	16	25
y	1.25	0.3125	0.2

34.

x	2.7	6.1	11.2
y	1,080	2,440	4,480

35.

x	4.4	7.8	11.2
y	0.176	0.312	0.448

36.

x	3.9	7.3	12.4
y	0.01026	0.00548	0.00323

H.   Solve the following:

37.   If j varies inversely as c, and $j = 7.5$ when $c = 5$, find j when $c = 13$.

38.   If I varies inversely as R, and $I = 2$ when $R = 1500$, find I when $R = 200$.

✔39.   If V varies directly as I, and V $= 24$ when $I = 3$, find V when $I = 5$.

Solve the following problems.

✔40.   Sales tax varies directly with the cost of a taxable item. If the sales tax on a \$350 item is \$21, find the sales tax on a \$4,500 item.

✔41.   The amount of pollution entering the atmosphere varies directly as the number of people living in the area. If 60,000 people cause 42,600 tons of pollutants, how many tons entered the atmosphere of Lexington? (Assume a population of 200,000).

✔42.   The profit realized by a textbook company varies directly as the number of algebra books sold. If the company makes a profit of \$4000 on the sale of 250 algebra books, what is the profit when the company sells 5000 algebra books?

43. The current in a simple electrical circuit varies inversely as the resistance. If the current is 20 amps when the resistance is 5 ohms, find the current when the resistance is 8 ohms.

44. The weight of an object on or above the surface of the earth varies inversely with the square of the distance of the object to the center of the earth. If an object weighs 1000 pounds at a distance of 5000 miles from the center of the earth, how much would it weigh 10,000 miles from the center of the earth?

45. The volume of gas varies inversely as the pressure and directly as temperature. If a gas occupies a volume of 1.5 liters at 346 K and a pressure of 18 Newtons per square centimeter, find the approximate volume at 340 K and a pressure of 24 Newtons per square centimeter.

46. Sketch the graphs of $y = x^5$ and $y = 5^x$, then compare and contrast them.

47. Sketch the graphs of $y = x^4$ and $y = 4^x$, then compare and contrast them.

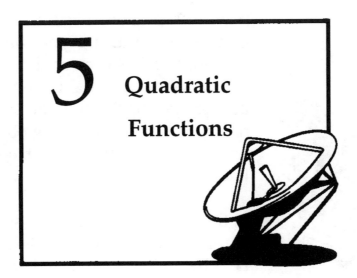

5 Quadratic Functions

Upon successful completion of this unit you should be able to:

1. Graph quadratic functions identifying vertices and intercepts;

2. Identify a quadratic function expressed in equation form;

3. Solve quadratic equations graphically, numerically, and algebraically;

4. Identify effects of the coefficient of x^2 and the constant term of a quadratic function on the graph of the parabola;

5. Write quadratic equations that represent real-life applications; and

6. Use the Pythagorean Theorem and the square root function where appropriate.

Throughout this book you have studied relationships algebraically, numerically, and graphically. In Unit 2 you studied linear functions. It has been said, however, that nature abhors a straight line. Consider the part of the world not made by man. Can you name 2 naturally occurring objects that are linear? And 2 man-made objects that are linear?

In this unit you will study a specific non-linear function, the quadratic function. Let's begin with "Discs and Squares."

Discs & Squares

In this experiment you will investigate the relationship between the length of the side of a square and the largest number of discs that can fit into the square without overlapping. You will see that the number of discs required *depends* on the length of the side of the square.

Equipment needed:
      Ruler
      Colored discs or pennies (or other material supplied by your instructor)
      Squares (provided on following pages)
      Graph paper
      Graphing calculator

Measure the length of one of the side of the squares on pages 2 and 3, record the length, then fill the square with as many discs as possible. Count and record the number of discs needed. This process should be repeated for each square. Enter your results in the table.

Length of Side						
Number of Discs						

The independent variable represents _____. Call it *s*.

The dependent variable represents _____. Call it *D(s)*.

Transfer the data collected to the table below. In addition to the data collected, add an additional entry which is the result of the following question: If you have a square of side length 0, how many discs will fill the square?

s							
D(s)							

On your graph paper, draw a rectangular coordinate system. Label the axes and show your scale. Plot the data points. If your points were connected with a smooth curve, would the curve be a line? Connect your points with a smooth curve to test your conjecture.

Once your group has graphed and connected the points on graph paper, plot the points using the graphing calculator. Be careful to choose an appropriate window. Once you have viewed the points, change the window several times. Are the points linear?

Parabolas

The curve your group obtained through careful measuring and counting is part of a curve which belongs to a family of curves called parabolas. Parabolas occur in nature in many different settings, and are useful in astronomy, optics, automobile manufacturing, artillery, and telecommunications. In this unit you will have the opportunity to explore parabolas.

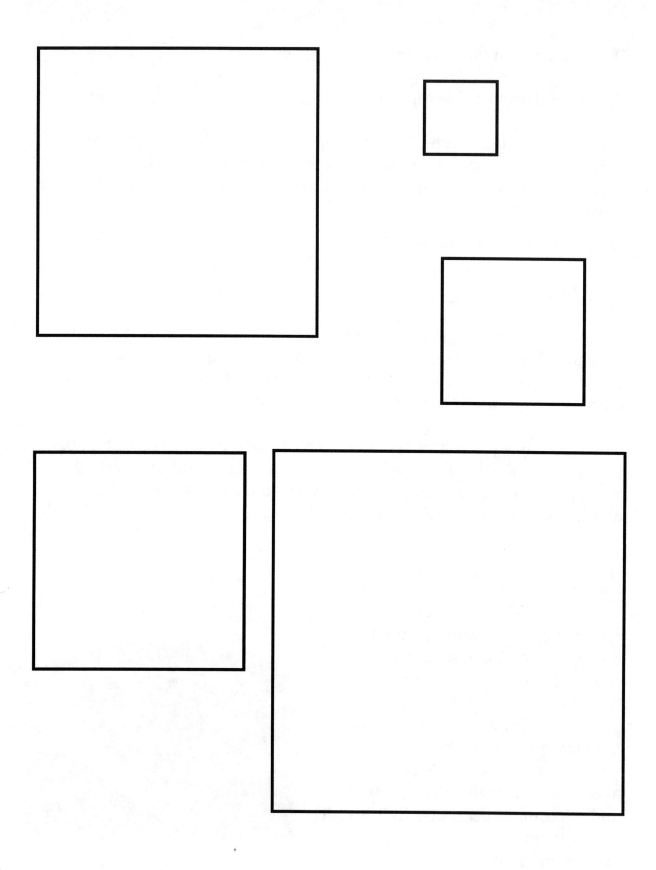

The simplest function whose graph is a parabola is $f(x) = x^2$. To explore the properties of this parabola, you will carefully graph points that are solutions to the equation, $y = x^2$.

For each x-value, calculate the y-value and enter it in the table. Examine the table for any patterns. Plot the pairs of values on the coordinate system on the next page using the scale given, and connect your points with a smooth curve.

x	-3.5	-3	-2.5	-2	-1.5	-1	-0.75	-0.5	-0.25	0
$y = f(x)$										

x	3.5	3	2.5	2	1.5	1	0.75	0.5	0.25
$y = f(x)$									

1. Compare and contrast $f(2)$ and $f(-2)$. $f(3)$ and $f(-3)$.

2. What is the smallest value of $f(x)$ in the table?

3. What is the lowest point on the graph?

4. What is the equation of the vertical line that passes through the lowest point on the graph?

5. Fold the graph along this vertical line.

6. Do the points (2, 4) and (–2, 4) coincide? Do the points (3, 9) and (3, –9) coincide?

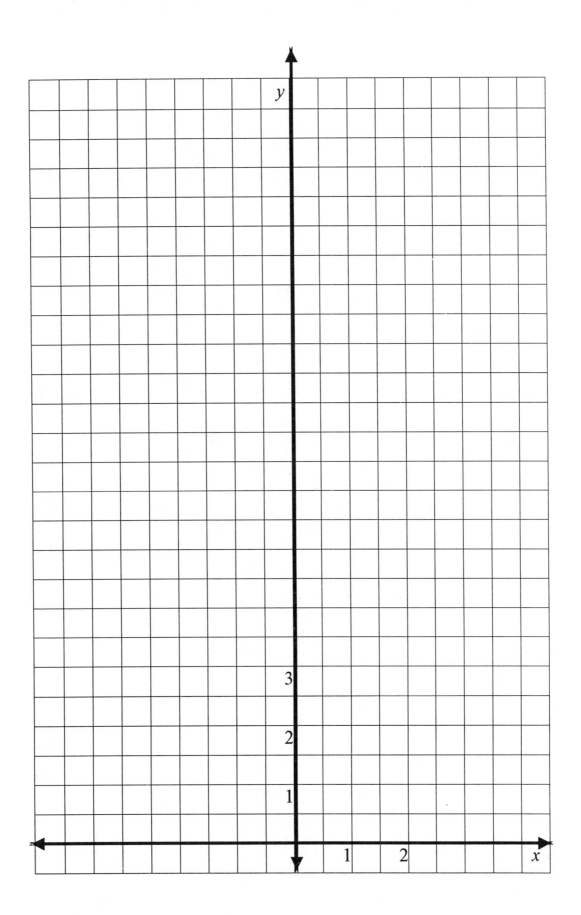

Your graph of $y = x^2$ should look similar to the graph shown below.

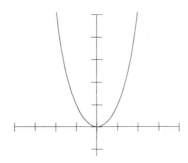

The lowest point on this graph is the **vertex** of the parabola. The vertical line that passes through the vertex is the **axis of symmetry** of the parabola. A parabola is symmetric about its axis. You looked at the axis of symmetry by folding. When you folded the parabola along its axis of symmetry, you found that the two parts of the parabola coincided. If a mirror were placed on the axis, each point on the parabola would have its image across the axis of symmetry. The mirror image is a point on the parabola. Look back at the table you just completed and note that the point (-3, 9) is the mirror image of the point (3, 9).

You Try It

Graph each of the following equations by calculating the specified values, plotting the points on graph paper and connecting them with a smooth curve. Identify the vertex and (using a different colored pen) draw in the axis of symmetry.

1. $y = x^2 - 3$
    $(-2, \quad), (-1, \quad), \left(\frac{-1}{2}, \quad\right), (0, \quad), \left(\frac{1}{2}, \quad\right), (1, \quad), (2, \quad)$

2. $y = (x - 3)^2$
    $(1, \quad), (2, \quad), (2.5, \quad), (3, \quad), (3.5, \quad), (4, \quad), (5, \quad)$

3. $y = (x + 1)^2 - 3$
    $(-3, \quad), (-2, \quad), (-1.5, \quad), (-1, \quad), (-0.5, \quad), (0, \quad)$

4. $y = 2(x - 3)^2 + 5$
    $(0.4, \quad), (2, \quad, (3, \quad), (4, \quad), (5.5, \quad), (0, \quad)$

Let's explore some other applications of parabolas.

Jumping Grasshoppers

Flights of leaping animals usually follow curved paths. The length of a leap of a grasshopper is 14 inches and the maximum height off the ground is 5 inches. Suppose this path is represented by the function, $h(d) = \frac{-5}{49}(d-7)^2 + 5$, where d is the distance in inches the grasshopper has jumped and $h(d)$ is the height of the path in inches of the grasshopper.

We can use the graphing calculator to examine point by point the path of the grasshopper. Using the STAT feature enter the values for all the integers from 0 to 14 for d in L1. Arrow over to highlight the L2. Enter (−5/49)(L1−7)²+5 or ⬚⬚(-)5÷49 ⬚⬚2nd1 −7⬚x²+5. 2nd1 is L1. The resulting list will contain the values calculated for $h(0)$ through $h(14)$.

L1	L2	L3
0	------	------
1		
2		
3		
4		
5		
6		

L2=.../49)(L1−7)²

How high is the grasshopper when it has traveled 5 inches horizontally? Where will the grasshopper reach its peak height? The grasshopper will have traveled how far horizontally when it is at height 0 again?

Now, set the WINDOW to $[-2, 16.8]_x$ and $[-2, 12]_y$. Plot the points.

To view the path of the grasshopper between the points you plotted, graph the quadratic function, $y = \frac{-5}{49}(x-7)^2 + 5$.

Trace the parabola to find the highest point of the grasshopper during its leap. Record this point.

Longest Home Run?

In 1919, Babe Ruth hit the longest home run ever recorded in a major league baseball game. We can approximate the path of Ruth's baseball by the function, $P(d) = -0.0017d^2 + d$ where d represents the horizontal distance in feet from home plate and $P(d)$ represents the vertical distance in feet of the baseball from the ground.

Here is the graph of the function created by a computer program. Compare this graph with one you graph on the graphing calculator.

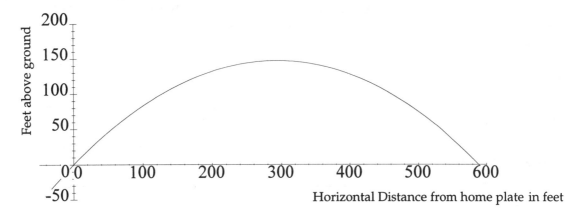

What are some questions you might want to answer using the graph of this record home run? Does this record still hold?

How high was the ball when it was 100 feet away from home plate? In function notation, this is asking what $P(100)$ equals.

What is $P(300)$?

Approximate the highest point in the path of the baseball. What was the horizontal distance of the baseball from home plate when it reached the highest point?

The Vertex of the Parabola

Clear all equations from your calculator and set the WINDOW to $[-4.7, 4.7]_x$ and $[-12, 12]_y$.

1. Complete the table below. (a) Factor the expression on the right hand side of each equation. (b) Graph and trace to determine the x-intercepts and vertex.

Parabola	Factored form	x-intercept(s)	Vertex
$y = x^2 - 4$			
$y = x^2 + 6x + 5$			
$y = x^2 - 2x - 3$			
$y = 4 - x^2$			
$y = -2x - x^2$			
$y = (x+1)^2$			
$y = (x+2)^2 - 4$			
$y = (x-2)^2 - 9$			

2. What is the connection between the x-coordinate of the x-intercepts and the x-coordinate of the vertex?

3. Can you develop a formula for the vertex?

4. Did you find any other patterns?

A quadratic function is any function of the form, $f(x) = ax^2 + bx + c$ where a, b, and c are real numbers and $a \neq 0$.

The graph of a quadratic function is a parabola. The x-coordinate of the vertex may be found using the formula $x = \dfrac{-b}{2a}$. The y-coordinate is then found by substituting the x-coordinate into the original function. Therefore, the vertex may be written, $\left(\dfrac{-b}{2a}, f\!\left(\dfrac{-b}{2a}\right) \right)$.

The equation of the axis of symmetry is $x = \dfrac{-b}{2a}$.

✎ For the *You Try It* on page 149, write the equations in problems 2, 3, and 4 in the form, $y = ax^2 + bx + c$. Using the formula for the vertex above, calculate its coordinates. Compare with the answers you obtained on page 149.

✎ For the **Longest Home Run?** problem, find the vertex using the formula above. Compare the vertex with your original answers to the questions on page 151:

(a) Approximately what was the highest point in the path of the baseball?

(b) What was the horizontal distance of the baseball from home plate when it reached the highest point?

Investigating Parabolas – Varying the Coefficients

In this activity you will examine parabolas by changing the coefficient of x^2. To begin the activity you should set the viewing window to $[-4.7, 4.7]_x$ and $[-12, 12]_y$.

Graph $y = x^2$ on your calculator, and then sketch it on the coordinate system below.

Continue your examination by graphing the following equations on your calculator: $y = 2x^2$ and $y = 8x^2$. Sketch each one before continuing. Create an equation of a parabola for which all the points of the graph (other than the point (0,0)) will lie between $y = 2x^2$ and $y = 8x^2$. Are there other equations whose graphs lie between these two graphs?

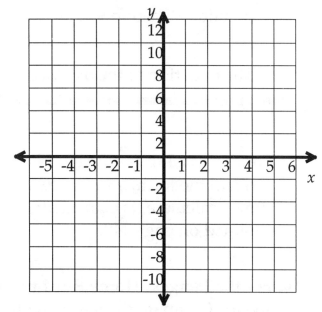

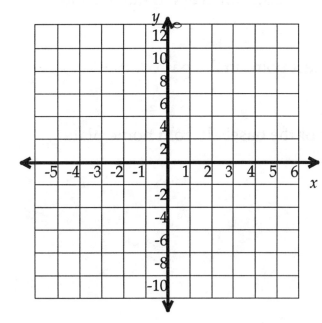

Discuss what the graph of each of the following should look like.

$$y = x^2$$
$$y = -4x^2$$
$$y = 4x^2$$
$$y = -3x^2$$

Graph the equations on the calculator, and then sketch them.

✎ Describe the graph of $y = ax^2$ when the coefficient a is positive?

✎ Describe the graph of $y = ax^2$ when the coefficient a is negative?

✎ Compare and contrast the graphs of $y = 2x^2$ and $y = 25x^2$.

✎ Compare and contrast the graphs of $y = -200x^2$ and $y = -3x^2$.

Now explore what happens as the constant term c changes.

Complete the following table for equations of the form, $y = ax^2 + bx + c$.

Equation	a	b	c
$y = x^2 + 2$			
$y = x^2 - 2$			
$y = x^2 + 6$			
$y = x^2 - 4$			

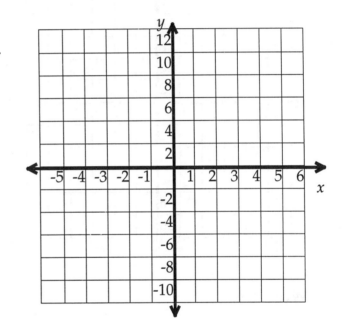

✎ What changes does your group expect to occur in the graph of $y = x^2$ as a result of adding or subtracting a positive constant?

Graph the equations from the table.

Choose values for c for the equation, $y = x^2 + 2x + c$. Complete the table, and then graph each equation.

Equation	a	b	c
$y = x^2 + 2x$			
$y = x^2 + 2x + \underline{\hspace{1cm}}$			
$y = x^2 + 2x + \underline{\hspace{1cm}}$			
$y = x^2 + 2x + \underline{\hspace{1cm}}$			

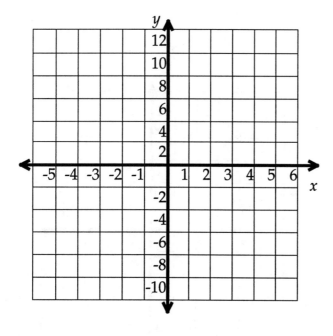

✎ If b is not equal to 0, does c still affect the graph in the same way?

Wrap Up

Given an equation written in the general form, $y = ax^2 + bx + c$, list the ways the coefficient of the x^2 term, a, affects the shape of the parabola.

_____

_____

List the ways the constant term, c, affects the shape of the parabola.

_____

_____

You Try It

Answer the following questions for the quadratic functions below.
  (a)   What is the vertex?
  (b)   What are the x-intercepts, if any?
  (c)   What are the y-intercepts?
  (d)   What is the equation for the axis of symmetry?

1. $f(x) = 2x^2 + 8x + 9$

2. $g(x) = -4(x-1)^2 + 3$

Determining Domain and Range of Quadratic Functions

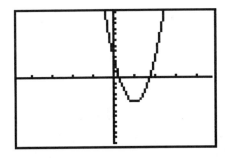
Suppose you are given the quadratic function, $f(x) = 8.2x^2 - 16.4x + 3.78$. Since all values of x produce real numbers for the function, $f(x) = 8.2x^2 - 16.4x + 3.78$, the domain of $f(x)$ is all real numbers. Another way to express this is to write $(-\infty, +\infty)$. All real numbers lie in this interval. Once you have determined the domain of the function, you can graph or visualize the function to find the range.

From the graph we see that $f(x)$ is a parabola opening upward. Since the lowest point on the graph occurs at the vertex when $x = 1$ and $f(1) = -4.42$, the range can be expressed as $[-4.42, \infty)$.

The chart below provides a review of the comparison of interval notation with the inequality representing the same values.

Interval Notation	Inequality
$[-2, 3]$	$-2 \le x \le 3$
$(-2, 3)$	$-2 < x < 3$
$(-\infty, 3]$	$-\infty < x \le 3$ or $x \le 3$
$[-2, \infty)$	$-2 \le x < \infty$ or $-2 \le x$

If an endpoint is to be included in an interval, use a bracket instead of a parenthesis. Always express domains and ranges by writing the smaller number first. When the interval notation includes infinity (∞), a parenthesis is always the correct symbol.

You Try It

Find the domain and range of each function.

1. $p(x) = -x^2 + 3.1x + 2.5$

2. $a(c) = c^2 - 2c - 4$

3. $f(x) = -2(x-3)^2 - 4.5$

Solving Quadratic Equations

We will now turn our attention to finding the zeros (solutions) of quadratic functions. To begin our study, let's start with an experiment that involves a topic with which we are all familiar, area. We will then move on to study free-falling bodies — an application from physics that is as old as the 1600's yet as timely as the twentieth century and beyond.

Experiment — The Area of the Dog Pen

Equipment needed:
      Centimeter graph paper – next page
      20 cm length of string per group

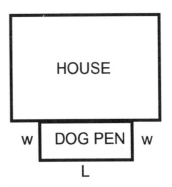

Suppose you have 20 ft. of fencing and are trying to build a ⊔-shaped dog pen attached to one side of your house. We will model this with a scale of 1 centimeter = 1 foot. Let w represent the width of the dog pen. $A(w)$ will represent the area of the enclosed region of the dog pen. What is the relationship between the width of the dog pen and the enclosed area?

1.   Use a string of fixed length 20 centimeters to form ⊔-shaped dog pens. Record values of the width w and length L in centimeters and the area $A(w)$ in square centimeters. Start with $w = 0$ and complete the table below.

w	0	1	2	3	4	5	6	7	8	9	10
L											
$A(w)$											

2.   Graph $A(w)$ versus w, and connect the points with a smooth curve.

3.   What value of w gives a maximum value for $A(w)$? What is the maximum area? Draw the shape of the ⊔-shaped dog pen with maximum area.

4.   Determine the domain and range of $A(w)$.

5.   Find a rule for the area $A(w)$ as a function of w.

6.   Find the values of w where $A(w) = 0$.

One-Centimeter Graph Paper

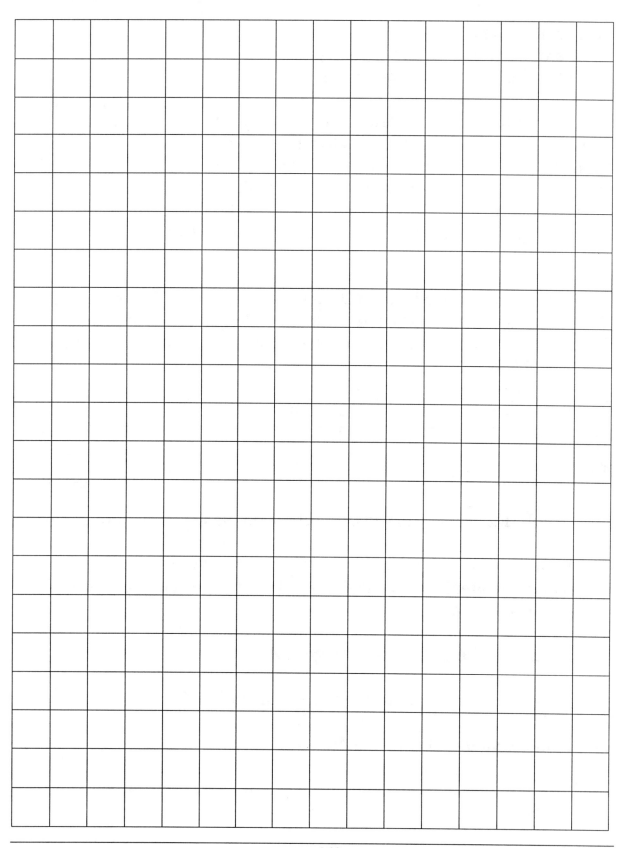

Falling Objects

Long ago Galileo studied the laws of falling bodies. He was an early philosopher who thought it necessary to test his theories by experiment. His conclusion was that all freely-falling bodies, regardless of their weight, fall the same distance in the same interval of time.

If you could drop a ball from the Tokyo Tower, which is 333 meters high, the height above the ground after one second would be 328 meters. After 2 seconds it would be 313 meters, etc. Here is a table which gives the height above the ground after t seconds.

Height in meters	328	313	289	255	211	157
Time in seconds	1	2	3	4	5	6

✎ Plot these heights versus times and look at the shape of the graph.

In many hot air balloon festivals, participants enter a race where balloon operators drop a marker to try to hit a target on the ground. Suppose the balloon is at a height of 1600 ft when the marker is dropped. Call this initial height h_0. Then the height in feet above the ground at any particular time is $h(t)$ given by

$$h(t) = h_0 - 16t^2$$

In the following investigation we will find how long it takes, neglecting air resistance, for a marker dropped from varying heights to hit the ground beginning with a height of 1600 ft.

1.   Graph the function $h(t) = -16t^2 + 1600$ on your graphing calculator.

   a)   Determine the value of t that makes $h(t) = 0$. This is the length of time it takes the marker to hit the ground.

   b)   Why would the balloon operator need to know this time?

   c)   Using the TABLE feature, how would you find the time?

   d)   Algebraically, to find this time, set $h = 0$ and solve for t.

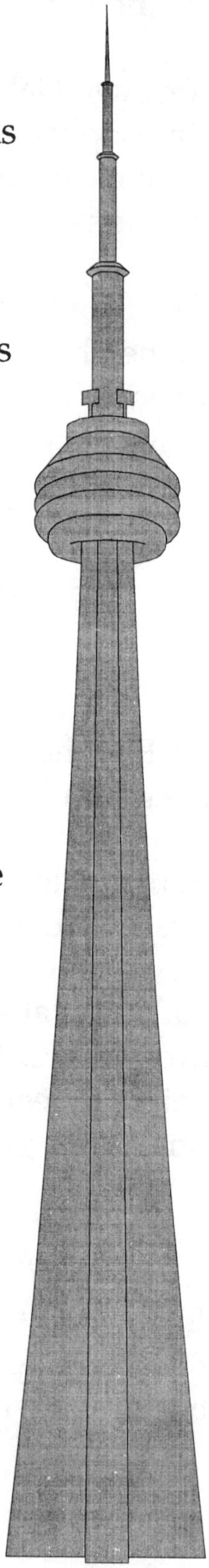

2.   Suppose the balloon were at $h_0 = 553$ m, the height of the CN Tower in Toronto. The CN Tower, currently the world's tallest building, is pictured here. Use $h(t) = h_0 - 4.9t^2$ which is adjusted for metric units.

   a)   Approximate the time necessary for the marker to hit the ground.

   b)   Describe the process of solving for the time above both graphically and algebraically.

Zeros

When we set an equation equal to zero and solve the equation we are finding the zeros of the function. This is the same as finding the x-intercepts of a graph. For example, solving the equation $x^2 - 4 = 0$ is the same as finding the zeros of the function, $f(x) = x^2 - 4$. One way to solve a quadratic equation is by using square roots.

You Try It

Solve using square roots, then confirm your solution graphically.

1. $x^2 = 100$

2. $x^2 = 454$

Another way is by factoring the equation and setting each factor equal to zero. Therefore, we also could have solved $x^2 = 100$ as follows:

$$
\begin{aligned}
x^2 &= 100 \\
x^2 - 100 &= 0 \\
(x - 10)(x + 10) &= 0 \\
x = 10 \text{ or } x &= -10
\end{aligned}
$$

Equations that factor easily can be solved quickly by factoring.

$$
\begin{aligned}
x^2 - 2x - 15 &= 0 \\
(x - 5)(x + 3) &= 0 \\
x - 5 = 0 \quad & \quad x + 3 = 0 \\
x = 5 \text{ or } & x = -3
\end{aligned}
$$

What do you do if the equation does not factor, or if you just can't find the factors? In these cases there is a formula called the **quadratic formula** to calculate the solutions of the equation.

The Quadratic Formula

The quadratic formula works for solving *any* quadratic equation, factorable or not. For the general quadratic equation $ax^2 + bx + c = 0$, the formula is stated in terms of the coefficients $a, b,$ and c:

$$x = \frac{-b \pm \sqrt{b^2 - 4ac}}{2a}$$

Note the square root in this formula. Solve the equation $x^2 - 4x - 2 = 0$ using the quadratic formula.

Fill in the blanks (...)

$$a = +1 \quad b = -4 \quad c = -2$$

$$x = \frac{-(\ldots) \pm \sqrt{(\ldots)^2 - 4(\ldots)(\ldots)}}{2(\ldots)}$$

$$x = \frac{(\ldots) \pm \sqrt{(\ldots)}}{(\ldots)}$$

$$x = (\ldots) \text{ and } (\ldots)$$

Graph the equation on your calculator to verify these zeros.

You Try It

Use the quadratic formula to find the zeros of each of the following functions.

1. $f(x) = 3x^2 + 4x - 7$

2. $g(x) = 4x^2 - 20x + 25$

3. $h(x) = x^2 + 4$

4. $j(x) = -2x^2 + 5x$

Again use your calculator to verify and interpret your results.

Freely-Falling Objects

A ball is tossed vertically into the air from the ground with a velocity of 112 ft. per second, and its height is measured at various times. The height of the ball can be described by the quadratic function:

$$height(t) = 112t - 16t^2$$

where *height* is measured in feet from the ground, and *t* is the number of seconds elapsed since the ball was thrown.

a) Graph the function on your graphing calculator and then make a sketch. Indicate the scales on the axes.

b) Trace to locate the point where *t* = 2.5. Interpret your answer.

c) At what time does the ball reach its maximum height? Use the TABLE feature to find the answer.

d) What is the maximum height?

e) When does the ball hit the ground?

f) How high is the ball after 2 seconds?

g) At what other time is the ball at this same height? Solve this algebraically as well as graphically.

h) At what time does the ball reach 100 feet? Did you get two answers? Explain why or why not.

Record your answers in the table on the next page.

Initial Height Given

Let's also consider the case where the ball from the previous example is thrown vertically upward from the top of a building 44 feet high. The height of the ball can be described by the quadratic function, $height(t) = 44 + 112t - 16t^2$.

Answer questions (a) - (h) for this case, comparing the answers step by step with those in the previous investigation.

	$h(t) = 112t - 16t^2$	$h(t) = 44 + 112t - 16t^2$	Observations
a.	Sketch on graph paper.		
b.			
c.			
d.			
e.			
f.			
g.			
h.			

Recap

In general the equation of motion for a freely-falling object is $s = \frac{1}{2}at^2 + v_0 t + s_0$. In this equation, s represents distance; a is acceleration; v_0 is initial velocity; s_0 is initial position; and t is time. If the only force acting on the body is gravity, then $a = -g$, where g is the acceleration due to gravity, 32 ft/sec^2 or 9.8 m/sec^2.

The Pythagorean Theorem

An application of square roots arises in using the Pythagorean Theorem which relates the lengths of the sides of a right triangle. A right triangle is a triangle where one angle is 90^0, a right angle. The side opposite the right angle is the hypotenuse and is the longest of the three sides. Each of the other sides is called a leg.

The Pythagorean Theorem

In a right triangle the sum of the squares of the lengths of the legs is the square of the length of the hypotenuse.

$$a^2 + b^2 = c^2$$

High Voltage Wires

Poles for high voltage electric wires are 250 m apart. On a cold winter day, the wires are observed to be horizontal. In Kentucky wires are taut at –20° C. During the summer heat, the wire is known to expand 0.75 m in length. To approximate the amount of the sag, let's assume the wire hangs in a "V" shape. Using this assumption, how great is the sag?

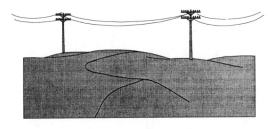

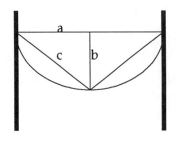

Bridge expansion   During the summer, heat expands a 2-mile long bridge by 2 feet in length. Assuming a bulge occurs straight up the middle, how high is the bulge? (The answer may surprise you. In reality, engineers design bridges with expansion spaces to avoid such buckling.)

The Box Problem

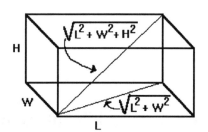

Measure the length L, width W and height H of a box in inches and record below.

$L =$ _____ $W =$ _____ $H =$ _____

Now measure the diagonal of the same box.

$d =$ _____

To find this diagonal algebraically requires the use of the Pythagorean Theorem in 3-dimensional space.

$d = \sqrt{L^2 + W^2 + H^2}$ (See figure above.)

Find the diagonal of the box algebraically. Compare your measurement for the diagonal and the algebraic approximation for the diagonal.

You Try It

Find the length of the diagonal of a box that is $9\frac{1}{4}$ inches long, $8\frac{3}{4}$ inches wide, and 2 inches deep.

Other Roots

In the previous problems you have used square roots. The radical symbol can indicate other roots as well. For example, the cube root of 27 is written $\sqrt[3]{27}$; the 3 to the left of the radical is the index of the radical.

For instance, suppose we need to find the length, l, of the edge of a cube with a volume of 64 cubic inches. We can solve the equation by taking the cube root of each side of the equation:

$$l^3 = 64$$
$$\sqrt[3]{l^3} = \sqrt[3]{64} \text{ or } (l^3)^{\frac{1}{3}} = 64^{\frac{1}{3}}$$
$$l = 4 \text{ in}$$

You Try It

The risk of death of a 40 – 49 year old man rises sharply above a certain threshold weight. This weight W (in pounds) is related to the man's height H (in inches) according to the function,

$$H(W) = 12.3\sqrt[3]{W}$$

What height corresponds to a threshold weight of 200 pounds for a man who is 41 years old?

 **Summary**

I. Quadratic Functions

♦ Form: $f(x) = ax^2 + bx + c$ — look for the squared variable
♦ The graph is a parabola. The constant term moves the parabola up or down c units depending on the sign of c.
♦ The coefficient of x^2 affects a parabola in the following ways:
>   If a is positive, the parabola opens upward.
>   If a is negative, the parabola opens downward.
>   The coefficient of x^2 also widens or narrows the parabola.
♦ Vertex: highest or lowest point on the parabola, $\left(\dfrac{-b}{2a}, f\left(\dfrac{-b}{2a}\right) \right)$
♦ Axis of Symmetry: mirror for the parabola, $x = \dfrac{-b}{2a}$
♦ When we set a function equal to zero and solve the equation we are finding the x-intercepts of a graph or the zeros of the function.

II. Quadratic equations can be solved by

♦ Using square roots;
♦ Factoring the equation and setting each factor equal to zero; or
♦ Using the Quadratic Formula, $x = \dfrac{-b \pm \sqrt{b^2 - 4ac}}{2a}$.

III. The Pythagorean Theorem states that in a right triangle the sum of the squares of the lengths of the legs is the square of the length of the hypotenuse.

$$a^2 + b^2 = c^2$$

IV. The equation for motion of a freely-falling object is $s = \dfrac{1}{2}at^2 + v_0 t + s_0$.

In this equation, s represents distance; a is acceleration due to gravity; v_0 is initial velocity; s_0 is initial position; and t is time.

ProCats

Memorandum #5   **Lexington, KY 40506**

TO: Team Members

FROM: Your Supervisor

The new souvenir shop, *The Cat's Meow,* is doing well! We are now considering adding a new line of shorts matching the T-shirts to be sold as sets. These shorts will cost $7.20 and sell for $14.95 as part of the set. The monthly cost of operating the shop fits the following function:

$$C(x) = 800 + (9.50 + 7.20)x + 0.05x^2$$

where the constant is the fixed cost (rent, utilities, etc.), the coefficient of x is attributed to cost of T-shirt and short material and the coefficient of x^2 could be labor cost. Management needs the team's assistance once again to decide if the addition of the short line is profitable.

The Analysis:

1. Construct a monthly revenue function.

2. Graph both the revenue and cost functions.

3. Algebraically, how many T-shirt and short sets must be sold for this line to be profitable? Confirm your answer graphically.

4. Construct the monthly profit function.

---

5. Estimate how many shirt and short sets must be sold for management to receive the greatest profit.

6. Graph the profit function and prepare a table of values to help you convince management to accept your team's recommendation.

The Report:

Write a report including the above information and conclusions and submit it to your supervisor. The report should explain your work, your analysis, any assumptions you made, and comments about the reliability of your results as well as your conclusions.

Problems for Practice

A. Graph each of the following.

1. $y = x^2 - 2x + 1$ 2. $y = x^2 - 2x - 2$

3. $y = x^2 - 3$ 4. $y = -3x^2 - 2x + 3$

5. $y = -\frac{1}{2}x^2 + 2$ 6. $y = -x^2 - 2x - 1$

B. Approximate the vertex of each of the following by tracing on your graphing calculator.

7. $y = x^2 - 2x + 7$ 8. $y = 2x^2 - 3$

9. $y = -x^2 - 2.4x - 1.5$ 10. $y = x^2 + 4x + 3$

C. Write an equation for the axis of symmetry for each of the equations in #7 - 10.

D. For each of the following, name the value for a, b, and c in $y = ax^2 + bx + c$.

11. $y = 2x^2 - 2x + 1$ 12. $y = x^2 + 2.34x + 5.38$

13. $x^2 - 2x - y + 7 = 0$ 14. $y = \sqrt{2}\,x - 7 + 2.8x^2$

15. $y = m + (12m/s)x - (4.9m/s^2)x^2$

16. $x^2 + 3x - 2y - 5 = 0$

E. Estimate the x-intercepts of each of the following parabolas by tracing on your graphing calculator. Verify algebraically.

17. $y = 2x^2 - 9x + 4$ 18. $y = x^2 - 13x + 42$

19. $y = x^2 + 20x + 96$ 20. $x^2 + 1.2x - 9.25 - y = 0$

F. Approximate to the nearest hundredth the x-intercepts of each of the following parabolas using the TABLE feature of your graphing calculator. Verify algebraically.

21. $y = 3x^2 + 13x - 10$ 22. $y = x^2 - 2x - 5$

23. $y = 6x + 8x^2 - 9$ 24. $y = -x + 3x^2 - 3$

G. Find the x-intercepts of each of the following parabolas by factoring.

25. $y = x^2 - 14x + 48$ 26. $y = x^2 - 14x + 13$

27. $y = 2x^2 - 9x + 4$ 28. $y = 5x^2 - 26x + 5$

29. $y = 3 - 26x - 9x^2$ 30. $y = 6x^2 - 7x - 10$

H. Determine the vertex for each of the parabolas in E, F, and G.

I. 31. Jim was hired last summer to be a human cannonball at Holiday World. The designers of the event determined that his path would follow the parabola: $y = 3x - 0.15x^2$ with the units measured in meters. Jon, his assistant, has a 3 square meter circular net to catch him when he lands. Jon places the net so that its center is 20 meters from the cannon.

a) Sketch the path Jim will travel.

b) Determine if Jim and Jon will remain friends, i.e., will the net catch Jim?

J. Choose the best answer.

32. The graph of the equation $y = 4x - 6$ is
    a) a parabola   b) a line

    c) a circle   d) a square

33. The graph of the equation $y = x^2 + 5x - 4$ is
a) a parabola
b) a line

c) a circle
d) a square

34. The graph of the equation $y = -(x - 2)^2 + 3$ is
a) a parabola
b) a line

c) a circle
d) a square

K. Sketch the graph of each of the following pairs of equations on the same coordinate system. Compare and contrast the two graphs and equations.

35.
$$y = -2x^2 + 3x + 1$$
$$y = -2x^2 + 3x - 1$$

36.
$$y = \tfrac{1}{2}x^2 - x + 2$$
$$y = \tfrac{1}{2}x^2 - x - 2$$

37.
$$y = 2x^2 - x - 2$$
$$y = 2x^2 - x + 3$$

38.
$$y = x^2 + x + 5$$
$$y = -x^2 - x - 5$$

L. Find the domain and range of each of the following.

39. $y = x^2 + 6x + 2$

40. $y = x^2 - 2x - 5$

41. $y = x^2 - 9x - 36$

42. $f(x) = -x^2 + 9x - 14$

43. $y = \tfrac{3}{4}x^2 - \tfrac{2}{3}x - 3$

44. $r(p) = 2.3p^2 - 3.5p + 1.2$

M. Identify each of the following as a linear function in x, a quadratic function in x, or other.

45.
$$y = 2x - 3$$
$$y = x^2 - 3x + 2$$
$$y = 2^x$$

46.
$$y^2 + x^2 = 64$$
$$x^2 - 78.5x + 245 = y$$
$$\pi x - 2.5\pi = y$$

47. The sum of a number x and the square of the number.
    The sum of two consecutive integers, x and $x + 1$.

48. The area of a square garden of side x.
    The perimeter of the garden.

N. Factor completely each of the following by factoring out the greatest common factor.

49. $10x^3 - 15x^2$ 50. $30x^3y^4 + 20x^4y^3$

51. $3x^3 - 3x^2 - 9x$ 52. $5x(a - 2) - 3y(a - 2)^2$

O. Factor the following binomials completely.

53. $49x^2 - 64y^2$ 54. $4x^2 - \dfrac{1}{4}$

55. $36x^2 + y^2$ 56. $x^4 - 16$

57. $(x - 2)^2 - 9$ 58. $(y + 4)^2 - 16$

P. Factor the following trinomials completely.

59. $x^2 - 7x + 12$ 60. $x^2 - x - 6$

61. $15 - 2x - x^2$ 62. $3x^2 - 3x - 6$

63. $x^2 + 3xy + 2y^2$ 64. $3x^2 - 6xy - 9y^2$

Q. Solve the following quadratic equations by factoring.

65. $2x^2 + x + 4 = 7$

66. $x^2 = 18 - 7x$

67. $x(4 - 5x) = -1$

68. $9x^2 - 12x = 0$

69. $0.02x + 0.01 = 0.15x^2$

70. $0.02x - 0.01 = -0.08x^2$

R. Solve the following problems about freely-falling objects.

71. Karin is on her fourth-floor hotel balcony pitching pebbles into a small pool below. Karin throws the pebble upward with an initial velocity of 40 feet per second from her position 50 feet above the water's surface. The pebble's height h in feet t seconds after the toss is $h = -16t^2 + 40t + 50$.

a) Determine the time at which the pebble is again 50 feet above the water.

b) Find the time at which the pebble strikes the water.

72. If an arrow is shot upward from a tower 144 feet above ground level its height in feet after t seconds is $h = 144 + 128t - 16t^2$. How long does it take the arrow to reach the ground?

73. Juan and Enrico are pitching and hitting baseballs. If Enrico hits a ball at a height three feet above the ground, its height h in feet after t seconds is $h = 3 + 75t - 16t^2$. Find the number of seconds it takes for the ball to hit the ground in the outfield.

74. In fielding practice, when Juan pitches a ball it is released at a height five feet above the ground. He hits it when it falls back to a height of four feet. If it is tossed with an initial velocity of 25 feet/sec its height h in feet t seconds after the toss is $h = 5 + 25t - 16t^2$. How many seconds elapse after the pitch before the ball is hit?

S. Solve the following quadratic equations by using square roots.

75. $x^2 = 81$

76. $x^2 = \dfrac{3}{4}$

77. $4x^2 - 45 = 0$

78. $(2x - 1)^2 = 25$

79. $2.1x^2 - 8.4 = 0$

80. $0.05x^2 = 0.015$

T. Use the quadratic formula to solve the quadratic equations.

81. $x^2 + 3x - 5 = 0$

82. $5x^2 + x = 1$

83. $2.2x^2 - 3.1x - 1.9 = 0$

84. $-x^2 + 10x = 25$

85. $-4x^2 = 9 - 12x$

86. $1.08x^2 + 6.21x = -4.7$

U. The Pythagorean Theorem states that $a^2 + b^2 = c^2$ for a right triangle labeled as shown. Use the Pythagorean Theorem to solve for the unknown in the following problems.

87. $a = 5$, $b = 12$, $c = ?$

88. $a = 24$, $b = 7$, $c = ?$

89. $a = ?$, $b = 4$, $c = \sqrt{30}$

90. $a = ?$, $b = 2\sqrt{6}$, $c = 8$

91. The longer leg of a right triangle is 12 centimeters. The hypotenuse is 3 centimeters less than twice the shorter leg. Find the length of the shorter leg.

92. The lengths in meters of the three sides of a right triangle are consecutive integers. Find the lengths of the three sides.

V. Application Problems

93. **Rate of a Braking Car** When a car brakes suddenly, its tires often leave skid marks. The length of the skid marks can be used to determine the velocity of the car when it braked using the formula $V = \sqrt{12d}$ where d is the length in feet of the skid marks and V is velocity in miles per hour. If a car leaves skid marks that are 293 feet long, how fast was the car traveling when it suddenly braked?

94. "On a clear day, you can see forever." Not really, but on a clear day the distance M (in miles) anyone can see from an unobstructed position H feet above sea level is given by the equation
$$M = 1.22\sqrt{H}$$
How far can a person see on a clear day from a height of 500 ft? Find the height at which a viewer must be positioned in order to see 20 miles?

95. **Surface Area** The surface area S of an animal is related to the length of the animal by the equation $L = \left(\dfrac{S}{k}\right)^{\frac{1}{2}}$ where k is a constant that depends on the animal. Find the length of an animal with a surface area of 973 square centimeters if $k = \dfrac{2}{7}$.

96. For the **Jumping Grasshoppers** problem on page 150, the quadratic functions was $h(d) = \dfrac{-5}{49}(d-7)^2 + 5$. Expand the right hand side of this equation, then arrange the terms so you can locate a, b, and c. Using the formula for the vertex, calculate its coordinates. Which coordinate matches the solution you reached when you located the highest point of the grasshopper during its leap?

⌘97. Diagonals

A diagonal is a line segment *other than a side* that joins two vertices of a polygon. Several polygons have been sketched below for your use. The

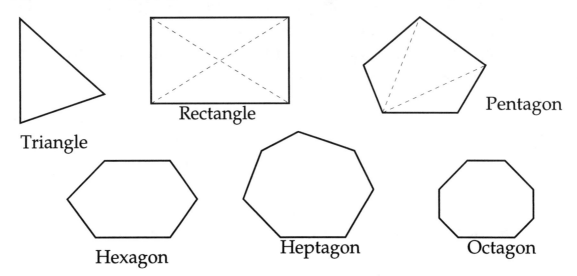

triangle has no diagonals. The rectangle which has four vertices has two diagonals. The diagonals are shown with dotted lines. The pentagon with five vertices has two of its diagonals drawn. Draw in the remaining diagonals for the pentagon, and then draw all diagonals for the remaining polygons. It is helpful to count the diagonals as you draw them rather than drawing them all and then counting them. Complete the table as you proceed. Observation of any patterns will help you complete this problem.

Polygons	Triangle	Rectangle	Pentagon	Hexagon	Heptagon	Octagon
Number of Vertices	3	4	5			
Number of Sides	3	4	5			
Number of Diagonals from Each Vertex	0	1				
Number of Diagonals	0	2				

Plot the results of the number of diagonals versus the number of sides. Can you connect the points with a straight line?

Write an equation that would allow you to determine the number of diagonals in a polygon with 100 sides.

⌘98. Was the temperature in December a function?

In Owensboro, Kentucky on December 28, 1994 the temperature was recorded hourly (not necessarily on the hour) using the twenty-four hour clock. The data collected from 0700 to 2030 is as follows:

T(0700)=30	T(1157)=55	T(1530)=60	T(1920)=40
T(0820)=39	T(1220)=57	T(1620)=58	T(2030)=37
T(0949)=46	T(1307)=59	T(1731)=55	
T(1043)=52	T(1425)=60	T(1827)=46	

All temperatures were recorded in degrees Fahrenheit. From the data collected, what values represent the domain, and what values represent the range?

Verify that T is a function of time by graphing the data below.

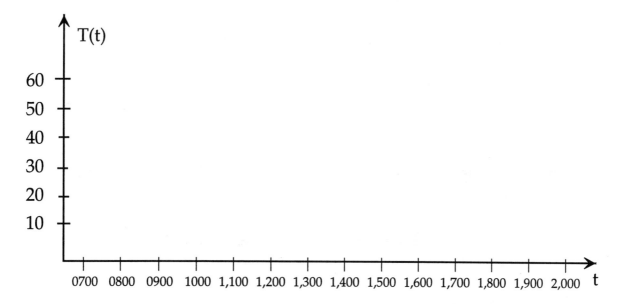

Connect the points on your graph with a smooth curve. If we look at time, all the values that can represent time are the domain. If we look at all the possible temperatures, we see the range. What are the domain and range as defined by the graph you've sketched?

Approximately what was the temperature at 0730, 1250, and 1745 that day?

⌘99. The Oil Tank Problem

Assume the number of barrels of oil remaining in an oil tank varies according to a quadratic function with the number of minutes that have elapsed since the valve was opened to drain the tank.

1.   If after 0.5, 1.5, 2.5, 3.5, 4, and 4.5 minutes since the valve was opened, the tank has 48.6, 29.4, 15, 5.4, 2.4, and 0.6 barrels remaining respectively, make a table with the data representing minutes elapsed and barrels remaining.

minutes elapsed						
barrels remaining						

2.   Next make a connected scatter plot containing the points tabulated. Notice the window settings.

3.   Explore the plot by tracing . What is the lowest point?

4.   A quadratic function closely approximating the table data is $y = 2.4x^2 - 24x + 60$. Graph this equation using $[-2, 16.8]_x$ by $[-5, 70]_y$ as the window dimensions.

5.   Using the graph determine how much oil was in the tank when the valve was opened.

6.   When will the tank have 5 barrels of oil remaining? Trace the graph or use the TABLE feature.

7.   When will the tank be empty?

8.   In the real world the number of barrels would never be negative. What is the lowest number of barrels the model predicts? Is this number reasonable?

9.   Why is a quadratic function more reasonable for this problem than a linear function?

6 Inequalities

Upon successful completion of this unit you should be able to:

1.   Solve linear and quadratic inequalities, graphically and algebraically;

2.   Interpret other inequalities graphically;

3.   Determine the domain of an inequality; and

4.   Solve applications involving linear inequalities.

As you study this unit there will be many situations that do not fall into the "cut and dried" category. Some problems have more than one solution; many have an array of answers. For example, when you fly on a plane you want the plane to fly at an altitude greater than the height of the mountain. When you go shopping, you know the most money you can spend. If you want to be in control of your college career, you must set goals for yourself as you move through your course of study. For instance, to balance a low score on an exam, you estimate the lowest score you need on the next test to get a good grade in the course.

Phrases such as "at least," "more than," and "between" all imply inequalities. When a company begins to manufacture a new part it must first know parameters for the part. Parameters are the acceptable ranges of size, weight, etc.

What is Acceptable?

Equipment needed:
    Parts
    Calipers

As part of Statistical Process Control, a manufacturer wants to determine if a machine is capable of making parts that meet customer specifications. In order to do this, a sample has been collected for your evaluation.

Measure each part and record its length.

Find the mean or average length of all the parts. Then determine the statistical range by subtracting the shortest length from the longest length.

Let $\bar{x}$ =mean length, a =shortest length, and b =longest length. Substitute the values into $a \leq \bar{x} \leq b$.

Explain just what this formula tells you. Is it possible to express this another way?   (Hint: $\bar{x} \pm c$)

But I gotta getta B!

1. Suppose you have one more test in history, and you
    want to have an average of at least 80 going into the
    final. Your test scores are 73, 90, and 81; each on a
    100-point test.

    a) Will a score of 82 earn you an average of at least
        80?

    b) Will 74 on the next test be sufficient?

    c) Write and solve an equation that will help you
        calculate the score you need for an average of
        80.

    d) Using an inequality, determine what range of scores will insure
        you earn an 80. Does your answer seem reasonable?

2. Your friend, Mike, is working on his grade also. Mike thinks that if
    he does well on the next test he could have a B average (80 - 89). His
    first three scores are 70, 58, and 92.

    a) What range of scores would give Mike an average of at least
        80?

    b) You think he will be lucky to get a C average (70 – 79), and you
        need to convince Mike you are right! Can you do this? Write
        an inequality and solve it.

3. It was the week before midterm and Alisha must decide whether to
    withdraw or to continue the course. She scored a 61 and a 67 on the
    first and second tests. She has 2 more 100-point tests to take, plus the
    200-point final exam. Discuss what average she must make on the
    two remaining tests and the final to:

    a) Make an A (90 – 100)?

b)   Make at least a B (80 – 89)?

c)   Make at least a C (70 – 79)?

d)   If you were Alisha's professor, what advice would you give her?

Developing Rules for Solving Inequalities

1.   Is 15 > 11?

   a)   Add the same positive number to each side. If you still have a true statement, record it.

   b)   Add the same negative number to each side of 15 > 11. Record this result.

   c)   Use your results to develop a rule (fill in the blanks):
        If $a < b$, then $a + c$ _____ $b + c$.

2.   Is –4 < 3?

   a)   Multiply each side by the same positive number. What happens?

   b)   Now multiply each side by the same negative number. Same results?

   c)   Discuss your observations.

   d)   Use your results to develop some rules. Remember that $c > 0$ means c is positive and $c < 0$ means c is negative.

   If $a < b$ and $c > 0$, then ac _____ bc.

   If $a < b$ and $c < 0$, then ac _____ bc.

Solving Inequalities

Summarizing your rules, you should observe that you solve an inequality exactly as you do an equation except if you multiply or divide by a negative number, you reverse the sense of the inequality (< to > or > to <). Some students avoid multiplying by a negative number by adding or subtracting terms so the coefficient of x is a positive number.

$$-3x < 12 \qquad \Leftrightarrow \qquad -3x + 3x - 12 < 12 + 3x - 12$$

$$\frac{-3x}{-3} > \frac{12}{-3} \qquad\qquad -12 < 3x$$

$$\qquad\qquad\qquad\qquad \frac{-12}{3} < \frac{3x}{3}$$

$$x > -4 \qquad\qquad\qquad -4 < x$$

These are the same answers! In interval notation both answers are represented by $(-4, \infty)$. Below is the line graph. Note the open circle above the -4 which indicates -4 is not part of the solution.

You Try It

Solve the following, sketch the line graph, and give the interval notation.

1. $4x + 9 \geq -15$

2. $-4x + 9 > -15$

3. $-5(x - 8) > 2(4x + 9)$

4. $\dfrac{(x-5)}{2} - \dfrac{(3x-4)}{3} \le -12$

5. $\dfrac{2}{5}(3x-4) \ge -2(x+6)$

6. Determine if your line graphs are reasonable.

✎ Complete the following sentence: If the coefficient of the variable is negative, then in the last step you must ....

Money, Money, Money

Eric has \$2000 to invest for a year. Find the range of annual simple interest rates that would result in interest income ranging from \$185 to \$225, inclusive. The formula for simple interest is

$$\text{Interest} = \text{Principal} \times \text{Rate} \times \text{Time}$$
$$I = p \cdot r \cdot t$$

If we let x be the unknown rate, then $I = 2000 \cdot x \cdot 1 = 2000x$. The inequality we need is

$$185 \le 2000x \le 225$$

Solving for x, then writing the solution with percentages, we obtain

$$0.0925 \le x \le 0.1125$$
$$9.25\% \le x \le 11.25\%$$

So the range of interest rates that would result in interest income ranging from \$185 to \$225, inclusive, would be interest rates from 9.25% to 11.25%.

You Try It

Selena just called a broker; she won the Pick 5 lottery and decided to invest her $100,000 for a retirement nest-egg. Her broker asked how much money she would like to accumulate over the next 25 years in simple interest alone. She had no idea and thought between $150,000 and a quarter of a million dollars would be great. Find the range of annual simple interest rates needed for these results.

1.   Write an inequality.

2.   Solve the inequality.

Compound Interest

When an initial deposit of P_o is placed in an investment paying an annual interest rate of r, compounded n times a year, then the total value, P_t, of the investment t years later is given by the formula

$$P_t = P_o\left(1 + \frac{r}{n}\right)^{nt}$$

If Selena invests her money in a CD paying 5.46% compounded annually, how much money will Selena have at the end of 25 years?

P_o is Selena's $100,000, r is 0.0546, and t is 25. So

$$P_t = 100,000\left(1 + \frac{0.0546}{4}\right)^{100}$$

Will Selena have met her goal of between $150,000 and $250,000 in interest?

✎ Use your graphing calculator to solve this problem. The best interest rate Selena can find is 5.74% compounded quarterly. During what range of years will Selena have between $150,000 and $250,000 in interest?

Millionaires

Selena has decided she wants to be a millionaire. If Selena found a high yield stock that pays 7.5% four times a year, when will Selena be a millionaire?

The equation we are interested in is

$$1,000,000 = 100,000\left(1 + \frac{.075}{4}\right)^{4t}$$

However, we have no algebraic methods to solve this problem at this time. With your graphing calculator, you can help Selena with her goal.

Subtract 1,000,000 from each side to obtain zero on one side of the equation.

$$0 = 100,000\left(1 + \frac{0.075}{4}\right)^{4t} - 1,000,000$$

Early in the book, you learned that the x-intercept occurs when $y = 0$. So if we set the right hand side of this equation equal to y, and then graph the equation, we will find the solution for t.

Graph $y = 100,000\left(1 + \frac{0.075}{4}\right)^{4t} - 1,000,000$.

X=30.988067 Y=-2.466E-5

Using the TABLE feature will help you set an appropriate window. Remember we need to know when $y = 0$. The y-value is very close to zero when x is approximately 30.988. However, Selena will not become a millionaire until the next compounding date or 31 years.

If Selena wants to know when she will have more than $250,000, you can use a similar approach to this problem.

The exponential equation you need to use is $P_t = 100,000\left(1 + \dfrac{0.075}{4}\right)^{4t}$. The inequality you are interested in is $P_t > 250,000$. Why? If you graph both of these equations on the same axes, you will need to find the intersection of the two graphs. You also need to know what part of the exponential graph lies above the line $y = 250,000$ to answer Selena's question. Graph these equations on the window $[0, 18.8]_x$ by $[-10, 300000]_y$ and complete this problem.

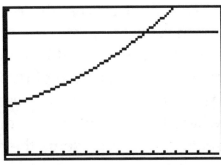

You Try It

Use your graphing calculator to solve the problems below.

1. On July 1 Roger sprayed his house for fleas, killing all but the 12 fleas in the vacuum cleaner. In Roger's house, the number of fleas in the house is a function of the number of days that have passed or $F(d) = 12(1.3)^d$. When will there be more than 150 fleas in the house?

2. Vicki invested $1,000 in an account paying 5.4% simple interest annually. When will Vicki have more than $2,000 in her account?

   a) What equation expresses the interest earned?

   b) What inequality will help answer the question?

   c) To solve the problem graphically, what two equations will you graph?

   d) When will Vicki have more than $2,000 in her account?

River Park Air

River Park Air sells tickets for two different prices. First-class seats are $1002, and coach seats are $584. The number of tickets sold varies from flight to flight. Ticket revenue is a function of two variables. The two variables can be represented as F and C where F represents the number of first-class seats, and C represents the number of coach seats.

1. Write a function for ticket revenue.

2. To pay expenses Pat, the C.E.O., needs to have at least $25,500 in ticket revenue every day. Write an inequality to express this situation.

3. Do ticket sales of 2 first-class seats and 40 coach seats satisfy your inequality?

4. Do ticket sales of 4 first-class seats and 37 coach seats satisfy your inequality?

5. If the 757 has seating for 40 first-class passengers and 210 coach passengers, find three other combinations of ticket sales that meet the goal.

6. What combination of tickets meets the revenue goal with the fewest total number of tickets sold? Explain why you think your answer is correct.

7. What other factors might be considered in answering number 6?

Graphing Inequalities

How would you graph the linear inequality $y \geq x - 1$? First, graph the line $y = x - 1$. This line divides the plane into the region above the line, the line, and the region below the line. Plot a point not on the line. Substitute the coordinates of the point into the inequality to determine if it is true. If it is true, then all points in that region are solutions for the inequality, and you shade that region. If it is false, you shade the other region. Why? These three steps are shown below.

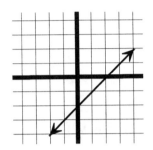 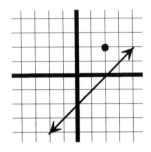 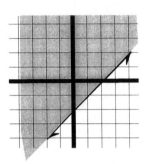

The three steps will allow you to graph any inequality for which you can graph the equation. If the sense of the inequality is strictly greater than or strictly less than, use a broken line or curve (- - - - - - -) instead of a solid line or curve.

You Try It

1. Graph the linear inequality $2x + y < 5$.

2. Graph the quadratic inequality $y \leq x^2 - 2x - 3$.

3. Graph the cubic inequality $y > x^3$.

✎ If a ball is tossed upward at 80 feet per second the distance in feet from the ground after t seconds is expressed by the equation

$$s = 80t - 16t^2$$

How many seconds will the ball be at least 52 feet above the ground? Use your graphing calculator and two equations to find the solution.

Determine the solution using the TABLE feature of your graphing calculator?

More on Domain

For each of the questions below, answer the question by explaining your response.

1.   Are any inequalities functions?

2.   Do all linear equations represent functions?

3.   Do all quadratic equations represent functions with x as the independent variable? What if y is the independent variable?

4.   Is the domain of all linear functions as well as all quadratic functions the set of all real numbers?

5.   Are the domains of $f(x) = 2x + 3$ and $g(x) = \sqrt{2x + 3}$ the same?

If you are still pondering the answer to number 5, did you graph both functions on your graphing calculator?

Investigating Domains Using Inequalities

Any x-values greater than or equal to -1.5 will return a real number in the function $f(x) = \sqrt{2x+3}$. Since the radicand (expression under the radical) must be greater than or equal to zero, you might write $2x+3 \geq 0$. Solving the inequality algebraically, you should obtain the same solution you determined graphically.

You Try It

Determine the domain for each of the following functions algebraically. Remember to look for restrictions on the independent variable.

1. $y = \sqrt{6-3x}$

2. $f(x) = \dfrac{3}{x}$

3. $r(p) = \sqrt{4p-6}$

4. $g(x) = \sqrt[3]{x-1}$

5. $f(x) = \dfrac{1}{x-4}$

6. $t(x) = \dfrac{5}{\sqrt{x-4}}$

7. $y = \dfrac{1}{\sqrt{x^2-3x-4}}$ (Hint: Use all the tools you have available!)

 Summary

I.   Linear inequalities can be solved algebraically using the following rules:
   ◆ If $a < b$, then $a + c < b + c$.
   ◆ If $a < b$ and $c > 0$, then $ac < bc$; or
   ◆ If $a < b$ and $c < 0$, then $ac > bc$.

II.   Inequalities can be solved graphically using the following three steps:
   ◆ Graph the function.
   ◆ Plot a point not on the function.
   ◆ Substitute the coordinates of the point into the inequality to determine if it is true. If it is true, then all points in that region are solutions for the inequality, and you shade that region. If it is false, you shade the other region.

III.   A graphing calculator allows the solution of non-linear inequalities graphically.

IV.   Applications of inequalities occur every day.

Final Note to the Student: If you feel there are a lot of unanswered questions relating to the material in this unit and the text, you are correct. If you remember the title, *An Introduction to Functions Through Applications*, you should expect to deal more with these topics in your next mathematics course. You should, however, be well-prepared to investigate the next level in your study of mathematics. We and your professor wish you well, and hope you will let your professor know how you are progressing in the study of mathematics.

ProCats

Memorandum #6   **Lexington, KY 40506**

TO: Team Members

FROM: Booster Club

The big game is coming up after this already tough season. Someone heard
the head coach and his two assistants in a big argument about the
strategy for the game.

The head coach said, "We must have a score of at least 88 points, and we
can't count on any foul shots being good from the team. Just look at
their record."

The first assistant coach moaned, "Tell them to go for the field goal. We
know that our guys can't score three-pointers against this hot-shot
team, and I agree our foul shot record stinks! I say, make at least 42
field goals, then we could win it."

The second assistant declared, "If we can get at least 34 good shots, field
goals and three-pointers combined, we might have a chance. I agree
about the foul shots. Forget them!"

How are you going to satisfy all three coaches and win the big game? If
you know what is good for you, you will deal with this! Otherwise,
it will be 100 laps around the court, and no hope for the
championship.

Good luck!

Problems for Practice

A. Solve each inequality for x, then graph (all but number 7).

1. $5x + 3 \geq 10$
2. $2(x - 5) < 4$

3. $5(x - 2) > -3(x + 5)$
4. $-3(2x - 12) \leq -41$

5. $-7x < -42$
6. $7x < -42$

7. $-7ax + b < c$
8. $4 < -2x + 1 < 8$

9. $5 < 3x - 4 \leq 18$
10. $0.2x + 3 \leq 0.92$

11. $-0.42x - 7 \geq 0.9$
12. $0.3x - 6 \geq 0.014$

13. $0.6x - 5 \leq 0.34(x + 7)$
14. $-3 < -x - 1.5 < 17$

15. $\frac{-2}{3}x \geq 50$
16. $-2 \leq \frac{4x - 6}{3} < 12$

17. $\frac{x + 5}{2} > \frac{x}{4} - \frac{x}{2}$
18. $\frac{2(x - 1)}{3} \leq \frac{x - 5}{3}$

19. $9 < \frac{3x - 2}{-5} \leq 12$
20. $-8 \leq \frac{3 - 2x}{6} < 15$

B. Solve the following problems.

21. Jack Terry has \$12,000 to invest. He needs to make an investment that will yield at least \$650 in 18 months. Find the lowest rate of interest that he can consider to meet his needs. His friend, Martin, thinks \$650 is too low, and he is determined to make \$1000. What is the lowest rate of interest Martin can consider?

22. For x such that $\frac{6}{5} \leq x \leq \frac{9}{5}$, find possible values for y, where $y = 5x - 7$. Solve this problem algebraically and graphically.

23. If $a < b$ and $ab > 0$, discuss all possibilities when you compare $\frac{1}{a}$ to $\frac{1}{b}$.

24. Matt has three more nickels than dimes and four times as many quarters as dimes. If Matt must have at least $4.60 to go to the grill, what is the fewest number of dimes that he must have? Interpret your answer.

C. Find three ordered pairs that satisfy each of the inequalities. Graph the solution sets.

25. $3x - y \geq 8$

26. $y > x^2 - 4$

27. $y \leq -x^2 + 3$

D. Find two ordered pairs that satisfy the systems of inequalities. Graph each of the following pairs on the same coordinate plane.

28. $\begin{array}{l} x + y > 3 \\ x - y < 4 \end{array}$

29. $\begin{array}{l} 3x - y < 9 \\ 2x + y > -3 \end{array}$

30. $\begin{array}{l} y < 4x + 1 \\ y > x^2 - 2 \end{array}$

31. $\begin{array}{l} y \geq 2x^2 - 3 \\ y \leq \frac{1}{2}x^2 \end{array}$

32. Graph the following four inequalities on the coordinate system.

$y \leq 1$
$x \leq 2$
$y \geq -2$
$x \geq -1$

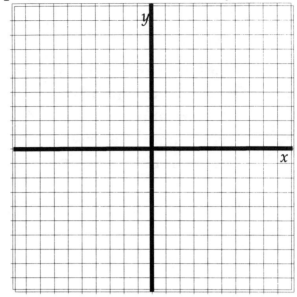

a)   Determine the area and perimeter of the figure.
b)   Write the equation for each diagonal.

33. Graph the following four inequalities on the coordinate system.

$y \leq 3x + 2$
$2y - 6x \geq -4$
$3y + x \leq 6$
$6y \geq -2x - 12$

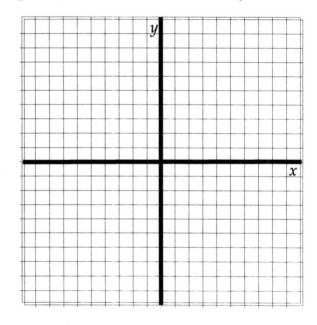

E. Determine the domain of each of the following functions.

34. $y = 6x - 12$

35. $y = \sqrt{6x - 12}$

36. $y = \dfrac{1}{6x - 12}$

37. $y = \dfrac{1}{\sqrt{6x - 12}}$

38. Are any two of the domains in numbers 34 – 37 exactly the same? If so, did you graph them to confirm your answer?

39. For what real numbers, x, is $f(x) = \sqrt{x^2 - \dfrac{4}{9}}$ defined?

40. Compare and contrast the domains of $g(x) = \sqrt{x - \dfrac{4}{9}}$ and $f(x) = \sqrt{x^2 - \dfrac{4}{9}}$.

F. Solve each of the following:

41. It costs \$3.00 to produce a beachball which sells for \$5.50. The monthly revenue is $R(x) = x(15 - 0.08x)$. How many beach balls must be sold each month to achieve a profit of at least \$400? (Profit = revenue – cost)

42. Jill can toss a ball upward at 64 feet per second. The distance s in feet of the ball from the ground after t seconds is expressed by the equation

$$s = 64t - 16t^2$$

a) How high is the ball after 1 second? After 2 seconds? After 4 seconds?

b) What will be the maximum height of the ball?

c) After how many seconds does the ball hit the ground?

d) For how many seconds will the ball be at least 32 feet high?

⌘43. The basic dimensions of any yacht competing for the America's Cup must satisfy the new rules. Suppose the following inequality states the dimension requirements with L, the length in meters; S, the sail area in square meters; and D, the displacement in cubic meters:

$$L + 1.25\sqrt{S} - 9.8\sqrt[3]{D} \leq 16.296$$

Determine whether the Katey Did It, with a length of 22.36 meters, a sail area of 265 square meters, and a displacement of 16.73 cubic meters satisfies this requirement.

⌘44. Is $(x+y)^2 \geq x^2 + y^2$? Consider combinations of positive and negative numbers for x and y.

⌘45. **Close Buys** sells boxes of 5.25 inch disks for $7.50, boxes of 3.5 inch disks for $12.50, and a back-up tape for $18.00. If f represents the number of boxes of 5.25 inch disks, t represents the number of boxes of 3.5 inch disks, and b represents the number of back-up tapes sold in one week, write an equation for Mark's weekly revenue as a function of f, t, and b.

a) Write an inequality to express a combination of sales that will provide a weekly revenue of at least $1200.00.

b) Find a combination of sales that will provide a weekly revenue of at least $1200.00.

c) Find a combination of sales that will provide a weekly revenue of less than $1000.00.

d) What combinations of sales yield weekly revenues between $1500 and $1750?

⌘46. Jon takes photos at college dances and other parties around town. He charges $5.50 for the each sitting. Jon sells two packages. A small package (S) sells for $9.50. The large package (L) is a better deal and sells for $14.50. What sales must Jon make to have an income of at least $250 at the dance?

⌘47. Write the inequalities represented by these graphs. (Note line ℓ in graph c intersects the axis at $(-\frac{3}{2}, 0)$ and $(0, -\frac{1}{2})$.

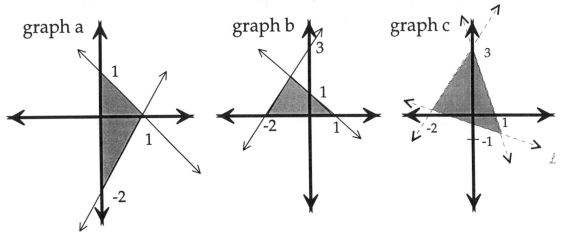

⌘48 Two Months Later at the Coffee Shop

Due to conflicts in Brazil the price of coffee beans is fluctuating wildly. The average cost per cup runs between 35 and 37 cents. The overhead is still $1200 a month, but the owner is not sure how to price her coffee. Her accountant suggests $1.35 since things can get worse. The owner is leaning towards $1.30 — she doesn't want to price herself out of business.

a)   Write a relation for the total monthly costs C to produce x cups of coffee costing between 35 and 37 cents.

b)   Find C(65). Is this a range of values?

c)   Write a relation for the revenue R for selling x cups of coffee if she decides the price per cup will be at least $1.30 but no more than $1.35.

d)   Graph both relations on the same coordinate system. Discuss the possibilities.

e)   The owner is a friend of yours and tells you she must break even. Tell her how much coffee she must sell a month if she prices on the low end and coffee costs 35 cents. Now do the same for the higher cost and higher selling price.

f)   If she decides to look at the worst case (highest cost, lowest price per cup), how many cups must she sell to make a profit of $1500 per month? $900 per month?

⌘✔49.   **Planting to Harvest**

The number of bushels of grain an acre of land will yield depends in part on how many pounds of seed per acre are planted. Seed is sold in 25-pound bags. From previous planting statistics, you find that if you plant 1.72 bags of seed per acre, you can harvest 22 bushels per acre, and if you plant 3.4 bags of seed per acre, you can harvest 40 bushels per acre. As you plant more seed per acre, the harvest will reach a maximum, then decrease. This happens because the young plants crowd each other out and compete for food and sunlight. Assume, therefore, that the number of bushels per acre you can harvest varies according to a quadratic function with the number of bags of seed per acre you plant.

1.   For the data given, ordered pairs (bags of seed planted, bushels harvested) can be written. For example, one such pair is (1.72, 22). A table with the number of bags of seed planted and the corresponding number of bushels per acre harvested is provided.

bags of seed planted	1.72	3.4	7.64	11.24	14.04	18.08
bushels harvested	22	40	70.15	78.29	73.71	43.03

2.   Make a scatter plot containing these six points.

3.   Explore the plot by tracing. What is the highest point on the plot?

4.   The standard form of a quadratic function is $y = ax^2 + bx + c$ where a, b, and c are constants. The graph of this equation is a parabola. Does the scatter plot look somewhat like a parabola?

5.   A quadratic function closely fitting the data is

$$y = -0.676x^2 + 14.8x - 2.04.$$

Graph this quadratic function on your calculator, using the window $[0, 20]_x$ by $[16, 84]_y$.

---

6. Trace the graph or use the TABLE feature on your calculator to determine how many bushels per acre you expect to get if 16 bags of seed per acre are planted.

7. Trace the graph or use the TABLE feature on the calculator to determine if it would be possible to get a harvest of 70 bushels per acre. Justify your answer.

8. Estimate how much should be planted per acre to get the maximum number of bushels per acre.

9. Now calculate how much should be planted per acre to get the maximum number of bushels per acre.

10. According to your graph is it possible to plant so many seed that there is no harvest of grain at all?

11. Set the quadratic function equal to zero, and solve algebraically.

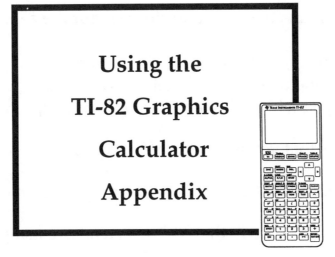

Using the

TI-82 Graphics

Calculator

Appendix

Contents

Check List:

These are some items to remember throughout the semester.

✓ Use the gray "(-)" for a negative sign and the blue "–" for subtraction. Although sometimes you will get an error message; sometimes you won't, but you will not have the answer you expected.

✓ Check to see that you are in the MODE you want.

✓ Turn off your STATPLOTS off before graphing a new function.

✓ Turn off all functions in the "Y=" screen before using STAT PLOT.

✓ Use the caret key, ⌐, for all exponents other than 2.

✓ Enclose all fractions in parentheses.

✓ Do not approximate fractions.

✓ Use your calculator efficiently.

✓ Remember the manual that came with your TI-82. It is a valuable resource.

✓ Replace your batteries when the number in the upper right hand corner is at 8 or 9 when increasing the contrast.

✓ Have fun and explore the capabilities of your TI-82. You really can't break it by using it.

✓ Don't leave your TI-82 to bake in your car in the summer or freeze in the winter.

Using the TI-82 Graphics Calculator

In each case an example will be followed by the appropriate keystrokes or instructions.

Finding the square of a number

Example: 4.2^2

Press ⁴4⁴ · ⁴2⁴ ⁴x²⁴ ⁴ENTER⁴. The correct answer is 17.64. Remember: The square affects only the number immediately preceding it so use parentheses to square a negative number.

Raising a number to a power

Example: 2.5^3

Press ⁴2⁴ · ⁴5⁴ ⁴^⁴ ⁴3⁴ ⁴ENTER⁴ . The ⁴^⁴ key is immediately above the ⁴÷⁴ key. The correct answer is 15.625.

Find the square root of a number

Example: $\sqrt{152.2}$

Press ⁴2nd⁴ ⁴x²⁴ ⁴1⁴ ⁴5⁴ ⁴2⁴ · ⁴2⁴ ⁴ENTER⁴. The answer on the screen should be 12.33693641.

Find the cube root of a number

Example: $\sqrt[3]{-27}$

Preferably, you would use the exponent, $\frac{1}{3}$, to obtain the cube root of –27. The keystrokes would be ⁴(⁴ ⁴(-)⁴ ⁴2⁴ ⁴7⁴ ⁴)⁴ ⁴^⁴ ⁴(⁴ ⁴1⁴ ⁴÷⁴ ⁴3⁴ ⁴)⁴ ⁴ENTER⁴.

Otherwise, press ⁴MATH⁴ to show the screen at the right. Press ⁴4⁴ for $\sqrt[3]{\ }$. Then enter (–27) and press ⁴ENTER⁴. The correct answer is –3.

Multiplying by a fraction.

Example: $\frac{9}{5} \cdot 10$

(1) Press ⎡(⎤ ⎡9⎤ ⎡÷⎤ ⎡5⎤ ⎡)⎤ ⎡1⎤ ⎡0⎤ ⎡ENTER⎤. The correct answer is 18.

(2) Press ⎡9⎤ ⎡÷⎤ ⎡5⎤ ⎡×⎤ ⎡1⎤ ⎡0⎤ ⎡ENTER⎤. This also gives the correct answer of 18.

Evaluating expressions with π

Example: $\frac{4}{3}\pi(3)^3$

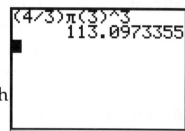

Press ⎡(⎤ ⎡4⎤ ⎡÷⎤ ⎡3⎤ ⎡)⎤ ⎡2nd⎤ ⎡^⎤ ⎡(⎤ ⎡3⎤ ⎡)⎤ ⎡^⎤ ⎡3⎤ ⎡ENTER⎤. The screen should match the screen to the right.

Plotting points

Example: Plot the points in the following table:

x	1	2	3	4
y	12	24	36	48

Press ⎡STAT⎤ to get the screen to the right.

Press ⎡1⎤ or press ⎡ENTER⎤ so you can edit the data in the list.

You should have a screen that matches the one to the right. If not, look up "clearing a list" (page A-6) before proceeding.

Now enter 1, 2, 3, and 4. Press enter after each number. To move to the next column press the right arrow ⎡▶⎤. In the second column enter 12, 24, 36, and 48. Press enter after each number.

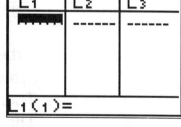

Plotting points (continued)

Now you are ready to plot the numbers. Press [2nd] [Y=] to access STAT PLOT. Your screen will look similar to this. Press [1] or [ENTER] to use **Plot 1**.

```
STAT PLOTS
1:Plot1...
   Off ⌐ L1 L2 .
2:Plot2...
   Off ⌐ L1 L2 □
3:Plot3...
   Off ⌐ L1 L2 □
4↓PlotsOff
```

Your screen should now look like the one to the right.

Press [ENTER] to turn Plot 1 on. **Type:** refers to the type of graph you want either a scatter plot ⌐, connected

```
Plot1
On Off
Type: ▦ ⌐ ⊞ ⅃⅃⅃
Xlist: L1 L2 L3 L4 L5 L6
Ylist: L1 L2 L3 L4 L5 L6
Mark: □ + .
```

scatter plot, box and whiskers graph, or a histogram. Most of the time we will use a scatter plot.

You just entered the x-values into L1 and the y-values into L2. The mark we will use this time is the square which is often easier to see. If your screen does not look like the one above, use your right and left arrows and press [ENTER] when the cursor is flashing on your choice.

Setting the WINDOW

Choosing a window is like deciding what marks we want to write on the graph paper. We need to decide what is the smallest x-value as well as the largest x-value. Then we decide how many tick marks we put on the x-axis. Then we repeat the process for the y-values. Press [WINDOW] enter these graphing instructions.

For our data, a possible window is the window to the right. Use the up and down arrows until the cursor is on the number you need to change so your screen matches the one at the right. Be careful to

```
WINDOW FORMAT
Xmin= -4.6
Xmax=4.6
Xscl=1
Ymin= -5
Ymax=50
Yscl=1
```

use the negative [(-)] and not the minus key [-]. Once your screen matches this one, press [GRAPH] to obtain your plotted points.

Setting the WINDOW (continued)

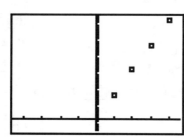

Can you find a WINDOW that gives you a graph you prefer?

Turning plots off

If you have been plotting points and want to use just the function graphing feature, press [2nd] [Y=] for STAT PLOT. Then press [4] and [ENTER] to turn all plots off.

Alternately, you could turn a plot off by entering the number of the plot you were using then backlighting the Off and pressing [ENTER].

Graphing a function

Example: Graph $f(x) = x^2$.

If you have been plotting points, turn your plots off by following the instructions under "turning plots off" above.

Press [Y=]. You can enter up to 10 different equations that will be graphed at the same time. To graph the function above, press the [X,T,θ] key which is in the second row, second column. Then press the [x²] key. Your screen should look like this.

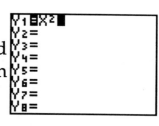

A WINDOW that we will frequently use is the ZDecimal WINDOW. To access the ZDecimal WINDOW quickly, press [ZOOM] [4]. [ZOOM] [4] simultaneously changes the WINDOW to the ZDecimal WINDOW below and graphs your function.

Tracing a function

Solve $3x - 9.6 = 0$ by graphing then tracing.

Graph the function $y = 3x - 9.6$ as described before. The solution of the equation is the x-intercept. We can find the x-intercept by tracing the graph.

With the graph of $y = 3x - 9.6$ on the screen, press [TRACE]. Use the right and left arrows to trace the graph until you reach the point where the y-coordinate is 0 as shown. So $3x - 9.6 = 0 \Rightarrow x = 3.2$.

Using the TABLE feature

Example: If $h(p) = p^2 + 2p + 3$, find $h(85)$ and $h(100)$.

Enter the function into [Y=] using x for the independent variable and y for the dependent variable. Press [2nd] [WINDOW] to access table setup (TblSet). Since we are interested in the point where $x = 85$, we might set the TblMin = 85. We also need to find the y-value when $x = 100$ so we might set the table change (ΔTbl) equal to 5. (If we chose 10, we would miss $x = 100$. Why?) Leave Indpnt and Depend on Auto.

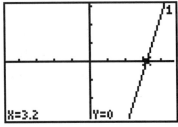

Finally, press [2nd] [GRAPH] for TABLE. The result should look like the screen at the right. So $h(85) = 7{,}398$ and $h(100) = 10{,}203$.

Clearing a list   Example:

```
 L₁    L₂    L₃
 1     12    ------
 2     24
 3     36
 4     48
 ------ ------
L₁(1)=1
```

To remove the values from list L1, use the up arrow △ until the L1 is backlit. Then press [CLEAR] and [ENTER]. Repeat this for any other lists you need to clear.

Evaluating a function for one value

Example: If $F(x) = 2x + 3.5$, what is $F(5.25)$?

Press [Y=], and enter $2x + 3.5$. Press [2nd][MODE] to return to the HOME screen. From the HOME screen, press [2nd][VARS] to access Y-VARS (y-variables). Choose option 1: Function..., then option 1: Y1. This process pastes Y1 on whatever screen you started from; in this case, the home screen.

Now, enter [(][5][.][2][5][)] and press [ENTER]. The resulting answer should be $F(5.25) = 14$.

Evaluating a function for several values

Example: If $F(x) = 2x + 3.5$, what are F(2), F(0), and F(7.37)?

Follow the instructions found in the first paragraph of "Evaluating a function for one value."

Then, press [(][2nd][(][2][,][0][,][7][.][3][7][2nd][)][)].
Press [ENTER]. Your screen should appear as the one at the right. The results are F(2) = 7.5, the y-intercept F(0) = 3.5, and F(18.24).

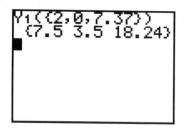

Trouble Shooting:

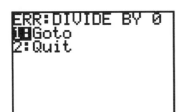

Your language does not match what the TI-82 expected. Press ENTER. Then closely examine what you told the TI-82 to do. For example, did you use " −" when you should have used the negative sign, "(-)"? Did you use parentheses correctly?

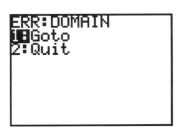

Division by zero is undefined; so look at your expression to see if an operation or variable makes the denominator of the fraction zero. This error does not occur during graphing.

This error occurs when you have made a mathematical error in a computation. For example, did you try to take the square root of a negative number? This error does not occur during graphing.

This error occurs when the number of entries in a list does not match the number of entries in another list being used to plot points. Check your lists for matching lengths. Check your STAT PLOTS to determine if two lists were unexpectedly paired.

This error occurs when you are trying to graph a function, but you cleared your lists and forgot to turn STAT PLOTS off.

Selected Answers

Unit 1

5. $H(4) \approx 5.7$ inches
   $H(44) \approx 44.2$ inches
   $H(444) \approx 444$ inches
7. \$48,131.70
9. a) $S\left(\frac{12.25}{2}\right) \approx 471.44$ sq. inches
10. a) 1.77
    b) 3
    c) 0.73
13. 1.93 sq. meters

Unit 2

3. a) For x: 9, 10, 11
      For y: 21, 23, 25
5. a) For x: $-7, -9, -11, -13$
      For y: $-17, -21, -25, -29$
15. $x = 4$
16. $x = \frac{33}{10}$
17. $p = \frac{-3}{2}$
18. $y = \frac{-2}{3}x + \frac{4}{3}$
19. $y = \frac{3}{4}x - 6$
21. $V = 40$
31. $m = \frac{1}{3}$
33. $m = 1$
35. $m = -4$
37. $m = \frac{-12}{5}$
39. $P(w) = 0.43w$
    $P(w^2) = 0.43w^2$
51. $\left(\frac{3}{2}, 0\right)$
53. $(7, 0)$
55. $\left(\frac{-3}{2}, 0\right)$
57. $(0, -4)$

59. $(0, 1)$
61. $(0, -5)$
63. $y = \frac{2}{3}x - \frac{4}{3}$
    $m = \frac{2}{3}$; y-intercept is $\left(0, \frac{-4}{3}\right)$
69. approximately 57 cm
77. a) $y = -3x + 17$
79. $y = \frac{-1}{6}x + 5.5$
83. $y = \frac{1}{2}x - 3$
85. The circumference is doubled.
87. a) $R(2) = 2$
    b) $R(-4) = 32$
    c) $s = \frac{28}{9}$
    d) $s = 0$
97. $x = 0$

Unit 3

1. $(3.1, 1.7)$
3. $(1.1, -1.4)$
5. $(2, 6)$
7. $(-2, -6)$
9. No solution
11. $(-25, -60)$
13. $(-8, -26)$
17. $\left(\frac{-1}{5}, \frac{22}{5}\right)$
19. $(-1, 2)$
21. $\left(\frac{10}{9}, \frac{-7}{6}\right)$
27. Neither
29. Perpendicular
31. $y = 3x + 11$
33. $y = -2x + 25$

Unit 4

1. Yes, D: $(-\infty, \infty)$
   R: $[-4, \infty)$
3. Yes, D: $[-6, 7]$
   R: $[-6.5, 6]$
5. No
7. No
9. Yes
11. Yes
17. 45.9 meters
25. $a = \dfrac{k}{\sqrt[3]{b}}$
27. Inverse, D: $(-\infty, 0) \cup (0, \infty)$
    R: $(-\infty, 0) \cup (0, \infty)$
29. Direct, D: $(-\infty, \infty)$
    R: $(-\infty, \infty)$
37. 2.88
39. 40

Unit 5

7. $(1, 6), x = 1$
9. $(-1.2, 0.06), x = -1.2$
11. $a = 2, b = -2, c = 1$
13. $a = 1, b = -2, c = 7$
17. $(0.5, 0), (4, 0)$
19. $(-8, 0), (-12, 0)$
21. $(-5, 0), (0.67, 0)$
23. $(-1.5, 0), (0.75, 0)$
27. $\left(\dfrac{1}{2}, 0\right), (4, 0)$
29. $\left(\dfrac{1}{9}, 0\right), (-3, 0)$
39. D: $(-\infty, \infty)$
    R: $[-3, \infty)$
43. D: $(-\infty, \infty)$
    R: $[-3.15, \infty)$
49. $5x^2(2x - 3)$
51. $3x(x^2 - x - 3)$

52. $(a - 2)[5x - 3y(a - 2)] =$
    $(a - 2)(5x - 3ya + 6y)$
53. $(7x - 8y)(7x + 8y)$
57. $(x - 5)(x + 1)$
61. $(5 + x)(3 - x)$
63. $(x + 2y)(x + y)$
65. $1, \dfrac{-3}{2}$
67. $1, \dfrac{-1}{5}$
69. $\dfrac{-1}{5}, \dfrac{1}{3}$
75. ± 9
79. ± 2
81. $1.19, -4.19$
83. $1.87, -0.46$
87. 13
89. $\sqrt{14}$

Unit 6

1. $x > \dfrac{7}{5}$
5. $x > 6$
9. $3 < x < \dfrac{22}{3}$
15. $x \le -75$
19. $\dfrac{-58}{3} \le x < \dfrac{-43}{3}$
35. $[2, \infty)$
37. $(2, \infty)$
39. $(-\infty, \dfrac{-2}{3}) \cup (\dfrac{2}{3}, \infty)$

Graphics Credits

Graphics Credits (continued)

Index

A

Acceleration due to gravity, 166
Area
    circle, 26
    square, 3, 19
Average, 184
Axis of symmetry, 149, 170
    equation, 153

B

Break-even point, 85

C

Calculator-Based Laboratory (CBL), 40, 106
Circumference
    circle, 4, 19
    ellipse, 17
CLEAR, 11
Compound interest, 189
Constant differences, 38
Coordinate system, 34
Cubic inequality
    solving, 193

D

Dependent variable, 39, 44, 54, 109
Diagonal, 168
Direct variation, 43
Domain, 108, 130, 194
    graphically, 111

E

Elimination, 88
Equation of a line, 63
Exponential
    decay, 132
    function, 127, 131
    growth, 129
Exponents, 15

F

Falling object, 161
Freely-falling object, 165, 170
    equation, 166
Function, 55, 63, 109, 119, 130
    applications, 5
    definition, 55, 63, 109
    exponential, 127, 131
    linear, 30
    notation, 4, 8, 12
    power, 126, 131
Functional notation, 4, 8, 12, 19

G

General equation, 48

H

Home screen, 10
Horizontal change, 50, 63
Horizontal lines, 62
Hypotenuse, 167